COMPUTATIONAL METHODS FOR RELIABILITY AND RISK ANALYSIS

SERIES ON QUALITY, RELIABILITY AND ENGINEERING STATISTICS

Series Editors: M. Xie (National University of Singapore)
T. Bendell (Nottingham Polytechnic)
A. P. Basu (University of Missouri)

Published

Vol. 1:   Software Reliability Modelling
   M. Xie

Vol. 2:   Recent Advances in Reliability and Quality Engineering
   H. Pham

Vol. 3:   Contributions to Hardware and Software Reliability
   P. K. Kapur, R. B. Garg and S. Kumar

Vol. 4:   Frontiers in Reliability
   A. P. Basu, S. K. Basu and S. Mukhopadhyay

Vol. 5:   System and Bayesian Reliability
   Y. Hayakawa, T. Irony and M. Xie

Vol. 6:   Multi-State System Reliability
   Assessment, Optimization and Applications
   A. Lisnianski and G. Levitin

Vol. 7:   Mathematical and Statistical Methods in Reliability
   B. H. Lindqvist and K. A. Doksum

Vol. 8:   Response Modeling Methodology: Empirical Modeling for Engineering
   and Science
   H. Shore

Vol. 9:   Reliability Modeling, Analysis and Optimization
   H. Pham

Vol. 10:   Modern Statistical and Mathematical Methods in Reliability
   A. Wilson, S. Keller-McNulty, Y. Armijo and N. Limnios

Vol. 11:   Life-Time Data: Statistical Models and Methods
   J. V. Deshpande and S. G. Purohit

Vol. 12:   Encyclopedia and Handbook of Process Capability Indices:
   A Comprehensive Exposition of Quality Control Measures
   W. L. Pearn and S. Kotz

Vol. 13:   An Introduction to the Basics of Reliability and Risk Analysis
   E. Zio

Vol. 14:   Computational Methods for Reliability and Risk Analysis
   E. Zio

Vol. 15:   Basics of Reliability and Risk Analysis: Worked Out Problems and
   Solutions
   E. Zio, P. Baraldi and F. Cadini

Series on Quality, Reliability and Engineering Statistics Vol. 14

COMPUTATIONAL METHODS FOR RELIABILITY AND RISK ANALYSIS

Enrico Zio

Polytechnic of Milan, Italy

World Scientific

NEW JERSEY · LONDON · SINGAPORE · BEIJING · SHANGHAI · HONG KONG · TAIPEI · CHENNAI

Published by

World Scientific Publishing Co. Pte. Ltd.
5 Toh Tuck Link, Singapore 596224
USA office: 27 Warren Street, Suite 401-402, Hackensack, NJ 07601
UK office: 57 Shelton Street, Covent Garden, London WC2H 9HE

Library of Congress Cataloging-in-Publication Data
Zio, Enrico.
   Computational methods for reliability and risk analysis / Enrico Zio.
      p. cm. -- (Series on quality, reliability & engineering statistics ; v. 14)
   Includes bibliographical references.
   ISBN-13 978-981-283-901-5
   ISBN-10 981-283-901-1
   1. Reliability (Engineering) -- Statistical methods. 2. Risk assessment -- Statistical methods.
   TA169 .Z598 2009

                                    2009277980

British Library Cataloguing-in-Publication Data
A catalogue record for this book is available from the British Library.

First published 2009
Reprinted 2011

Printed in Singapore.

To Sleeping Beauty, Snowhite and Saetta Mc Queen
To my wife Giorgia, author of my life book

Milano, 20 July 2008

Reliability and safety are fundamental attributes of any modern technological system. In practice, diverse types of protection barriers are placed as safeguards from the hazard posed by the system operation, within a *multiple-barrier* concept. These barriers are intended to protect the system from failures of any of its components, hardware, software, human and organizational.

Correspondingly, the reliability and risk analyses of a given system aim at the quantification of the probability of failure of the system itself and of its protective barriers.

A fundamental issue in these analyses is the uncertainty in the failure occurrences and consequences. For the objectives of system safety, this entails protecting the system beyond the uncertainties of its accidental scenarios.

One classical way to defend a system beyond the uncertainty of its failure scenarios has been to:

i)   identify the group of failure event sequences leading to credible *worst-case* accident scenarios $\{s^*\}$ (*design-basis accidents*),

ii) predict their consequences $\{x^*\}$ and

iii) accordingly design proper safety barriers for preventing such scenarios and for protecting from, and mitigating, their associated consequences.

Within this *structuralist, defense-in-depth* approach, safety margins against these scenarios are enforced through conservative regulations of system design and operation, under the creed that the identified worst-case, credible accidents would envelope all credible accidents for what regards the challenges and stresses posed onto the system and its protections. The underlying principle has been that if a system is designed to withstand all the worst-case credible accidents, then it is 'by definition' protected against any credible accident [1].

This approach has been the one classically undertaken, and in many technological instances it still is, to protect a system from the uncertainty of the unknown failure behaviours of its components, systems and structures, without directly quantifying it, so as to provide reasonable assurance that the system can be operated without undue risk. However, the practice of referring to "worst" cases implies subjectivity and arbitrariness in the definition of the accidental events, which may lead to the consideration of scenarios characterized by really catastrophic consequences, although highly unlikely. This may lead to the imposition of unnecessarily stringent regulatory burdens and thus excessive conservatism in the design and operation of the system and its protective barriers, with a penalization of the industry. This is particularly so for those industries, such as the nuclear, aerospace and process ones, in which accidents may lead to potentially large consequences.

For this reason, a more rational and quantitative approach has been pushed forward for the design, regulation and management of the safety of hazardous systems. This approach, initially motivated by the growing use of nuclear energy and by the growing investments in aerospace missions in the 1960s, stands on the principle of looking quantitatively also at the reliability of the accident-preventing and consequence-limiting protection systems which intervene in all potential accident scenarios, in principle with no longer any differentiation between credible and incredible, large and small accidents [2]. Initially, a number of studies were performed for investigating the merits of a quantitative approach based on probability for the treatment of the uncertainty associated with the occurrence and evolution of accident scenarios [3]. The findings of these studies motivated the first complete and full-scale probabilistic risk assessment of a nuclear power installation [4]. This extensive work showed that indeed the dominant contributors to risk need not be necessarily the design-basis accidents, a 'revolutionary' discovery undermining the fundamental creed underpinning the structuralist, defense-in-depth approach to safety [1].

Following these lines of thought, the probabilistic approach to risk analysis (PRA) has arisen as an effective way for analysing system safety, not limited only to the consideration of worst-case accident scenarios but extended to looking at all feasible scenarios and its related consequences, with the probability of occurrence of such scenarios becoming an additional key aspect to be quantified in order to rationally and quantitatively handle uncertainty [4-11]. From the view point of safety regulations, this has led to the introduction of new criteria which account for both the consequences of the scenarios and their probabilities of occurrence under a now *rationalist,* defense-in-depth approach. Within this approach to safety analysis and regulation, reliability engineering takes on a most relevant role in the assessment of the probability of occurrence of the accident scenarios.

In this book, a number of methods for computing the reliability and risk characteristics of complex technological systems are illustrated. The presentation of the theory behind the methods is of pedagogical nature, but supported with practical examples for a clearer understanding of how these methods can be applied in the field.

Chapter 1 introduces the basics of the Markov approach to system modeling for reliability and availability analysis. In this approach, the stochastic process of evolution of the system in time is described through the definition of the system states, the possible transitions among these states and their probabilities of occurrence. The various system states are defined in terms of the states of the components comprising the system. The components are not restricted to having only two possible states but rather may have a number of different states such as functioning, in standby, degraded, partially failed, completely failed, under maintenance, etc.; the various failure modes of a component may also be defined as states. The transitions between the states occur randomly in time, because caused by various mechanisms and activities such as failures, repairs, replacements and switching operations, which are random in nature. Under specified conditions, the stochastic process of the system evolution may be described as a so called Markov process which is mathematically described by a system of probability equations which can be solved analytically or numerically.

Chapter 2 gives a short introduction to the theory of Monte Carlo simulation for reliability and availability analysis. The presentation is kept at an intuitive and practical level. The Monte Carlo simulation method is shown to offer a powerful tool which can be of great value in the analysis of complex systems, due to its inherent capability of achieving a closer adherence to reality in the representation of the system stochastic behaviour. In general terms, it may be defined as a methodology for obtaining estimates of the solution of mathematical problems by means of random numbers. By random numbers we mean numbers obtained through a roulette-like machine of the kind utilized in the gambling casinos at the Montecarlo Principate: hence the name of the method. The random sampling of numbers was utilized in the past, well before the development of the present computers, by skillful scientists. The first example of use of what we now call Monte Carlo method seems to go back to the French naturalist Buffon (1707-88) who considered a set of parallel straight lines a distance D apart onto a plane and computed the probability P that a segment of length $L < D$ randomly positioned on the plane would intersect one of these lines. The theoretical expression he obtained was

$$P = \frac{L/D}{\pi/2}$$

Possibly not completely convinced about the correctness of his result, Buffon had the idea of checking the above expression by actually drawing parallel lines and repeatedly throwing a needle on the floor of his house to experimentally estimate the probability P as the ratio of the number of intersections to the total number of throws, thus acquiring the honour of being the inventor of the Monte Carlo method. It is interesting to mention that, later on, Laplace noticed that the Buffon's experiment represented a device for computing π just by throwing a needle on a floor with parallel lines. Successively other scientists used similar methods to solve integrals and probability problems. Eventually, the revival of the method seems to be ascribed to Fermi, von Neumann and Ulam in the course of the Manhattan Project during World War II. Back then, the Monte Carlo method provided the only option for solving the six-dimensional integral equations employed in designing shielding for nuclear devices. It was probably the first case in human history in which solutions based on trial and error were clearly too risky. Currently, Monte Carlo simulation seems to be the only method that can yield solutions to complex multi-dimensional problems. For about three decades it was used almost exclusively, and extensively, in nuclear technology. Presumably, the main reason for its use being limited to only nuclear applications was the lack of suitable computing power: indeed, the method is computer memory- and time-intensive. With the increasing availability of fast computers the application of the method becomes more and more feasible in the practice of various fields, including reliability and risk analysis.

Chapter 3 combines the modeling power of the Markov approach with the computing power of Monte Carlo simulation. This gives rise to the so called Markov Chain Monte Carlo techniques which offer an effective way for sampling from complicated probability distributions in high-dimensional spaces. This is useful in such tasks as image reconstruction, parameter identification, computing the equilibrium distribution and associated energy levels of statistical mechanics systems, inverse problem solving and more generally Bayesian posterior inference. Examples of application are provided with respect to the characterization of the failure and degradation behaviours of components and structures.

Chapter 4 illustrates the use of Genetic Algorithms within the area of RAMS (Reliability, Availability, Maintainability and Safety) optimization. The theory behind the operation of genetic algorithms is presented. The steps of the algorithm are sketched to some details for both the traditional breeding procedure as well as for more sophisticated breeding procedures. The necessity of affine transforming the fitness function, object of the optimization, is discussed in detail, together with the transformation itself. Finally, two examples of application are illustrated with regards to problems of reliability allocation and periodic inspection and maintenance. RAMS optimization is classically based on quantifying the effects that design and operation choices and testing and maintenance activities have on a number of system attributes like:

- $R(\underline{x})$ = System Reliability;

- $A(\underline{x})$ = System Availability ($U(\underline{x})$= system unavailability =$1 - A(\underline{x})$);

- $M(\underline{x})$ = System Maintainability, i.e. the unavailability contribution due to test and maintenance;

- $S(\underline{x})$ = System Safety, normally quantified in terms of the system risk measure Risk($\underline{x}$) (e.g. as assessed from a Probabilistic Risk Analysis);

where $\underline{x}$ represents the vector of the design, operation and maintenance decision variables. A quantitative model is used to asses how the design, operation and maintenance choices affect the system RAMS attributes and the involved costs ($C(\underline{x})$ = Cost required to implement the vector choice $\underline{x}$). Thus, the design, operation and maintenance optimization problem must be framed as a multiple criteria decision making problem where RAMS&C attributes act as the conflicting decision criteria with the respect to which optimization is sought and the relevant design and maintenance parameters (e.g. redundancy configuration, component failure rates, maintenance periodicities, testing frequencies) act as the decision variables $\underline{x}$. Then, the multiple criteria decision-making analysis aims at finding the appropriate choices of reliability design, testing and maintenance procedures that optimally balance the conflicting RAMS and Costs (RAMS&C) attributes. In this general view, the vector of the decision variables $\underline{x}$ encodes the parameters related to the inherent equipment reliability (e.g. per demand failure probability, failure rate, etc.) and to the system logic configuration (e.g. number of redundant trains, etc.), which define the system reliability allocation, and those relevant to testing and maintenance activities (test intervals, maintenance periodicities, renewal periods, maintenance effectiveness, mean repair times, allowed downtimes, etc...) which govern the system availability and maintainability characteristics.

Chapter 5 investigates the issues related to dependent failures and illustrates the approaches used to model their effects on system reliability. This is a quite crucial issue in reliability and risk analysis since in spite of the fact that all modern technological systems are highly redundant, they still fail because of dependent failures which can defeat the redundant system protective barriers and thus contribute significantly to risk; quantification of such contribution is thus necessary to avoid gross underestimation of risk.

Chapter 6 is devoted to the presentation of the concept of importance measure in reliability and risk analysis. From a broad perspective, importance measures aim at quantifying the contribution of components to the system performance, e.g. its reliability, availability or safety. For example, the calculation of importance measures is a relevant outcome of the Probabilistic Risk Assessment (PRA) of nuclear power plants which allows evaluating the relevance of components (or more generally, events) with respect to their impact on the risk measure of interest, usually the Core Damage Frequency (CDF) or the Large Early Release Frequency (LERF). In other system engineering applications, such as aerospace and transportation, the impact of components is considered on the system unreliability or, for renewal systems such as the manufacturing production and power generation ones, on the system unavailability. Information about the importance of the components constituting a system, with respect to its safety and availability, is of great practical aid to system designers and managers. Indeed, the identification of which components mostly determine the overall system behavior allows one to trace system bottlenecks and provides guidelines for effective actions of system improvement.

Chapter 7 provides some basic notions related to sensitivity and uncertainty analysis, in support to the analysis of the reliability and risk of complex systems under incomplete knowledge of their behavior. Indeed, as mentioned at the beginning, uncertainty is an unavoidable component affecting the behavior of systems and more so with respect to their failure limits. Thus, uncertainties arise in the values of the parameters and in the hypotheses on the structure of the models used to represent the system failure behavior. Such uncertainties propagate within the model used to compute the system reliability and risk, which become uncertain themselves. In spite of how much dedicated effort is put into improving the understanding of systems, components and processes through the collection of representative data, the appropriate characterization, representation, propagation and interpretation of uncertainty will remain a fundamental element of the reliability and risk analyses of any complex system. With respect to uncertainty, the final objective of reliability analysis and risk assessment is to produce insights in the analysis outcomes which can be meaningfully used by the decision makers. This entails that a number of topics be successfully addressed [12]:

- How to collect the information (e.g. in the form of expert judgment) and input it into the proper mathematical format.
- How to aggregate information from multiple, diverse sources into a single representation of uncertainty.
- How to propagate the uncertainty through the model so as to obtain the proper representation of the uncertainty in the output of the analysis.
- How to present and interpret the uncertainty results in a manner that is understandable and useful to decision makers.
- How to perform sensitivity analyses to provide insights with respect to which input uncertainties dominate the output uncertainties, so as to guide resources towards an effective uncertainty reduction.

In general, uncertainty can be considered essentially of two different types: randomness due to inherent variability in the system (i.e., in the population of outcomes of its stochastic process of behavior) and imprecision due to lack of knowledge and information on the system. The former type of uncertainty is often referred to as objective, aleatory, stochastic whereas the latter is often referred to as subjective, epistemic, state-of-knowledge [12,13]. Whereas epistemic uncertainty can be reduced by acquiring knowledge and information on the system, the aleatory uncertainty cannot and for this reason it is sometimes called irreducible uncertainty.

The distinction between aleatory and epistemic uncertainty plays a particularly important role in the risk assessment framework applied to complex engineered systems such as nuclear power plants. In the context of risk analysis, the aleatory uncertainty is related to the occurrence of the events which define the various possible accident scenarios whereas epistemic uncertainty arises from a lack of knowledge of fixed but poorly known parameter values entering the evaluation of the probabilities and consequences of the accident scenarios [12].

With respect to the treatment of uncertainty, in the current reliability analysis and risk assessment practice both types of uncertainties are represented by means of probability distributions [6]. Alternative representations based on different notions of uncertainty are being used and advocated in the context of reliability and risk analyses [12,14-16], questioning whether uncertainty can be represented by a single probability or whether imprecise (interval) probabilities are needed for providing a more general representation of uncertainty [17- 20]. It has also been questioned whether probability is limited to special cases of uncertainty regarding binary and precisely defined events only. Suggested alternatives for addressing these cases include fuzzy probability [21-23] and the concept of possibility [24-26]. Furthermore, probabilities have been criticised for not reflecting properly the weight of the evidence they are based on, as is done in evidence theory [27].

The issue of which framework is best suited for representing the different sources of uncertainty is still controversial and worth of further discussion. In the Chapter, the discussion is limited to the probabilistic representation of uncertainty, which is currently the most widely used in practice. A recent critical review of the alternative frameworks of representation of uncertainty is provided in [28], from the starting point of view that a full mathematical representation of uncertainty needs to comprise, amongst other features, clear interpretations of the underlying primitive terms, notions and concepts. The review shows that these interpretations can be formulated with varying degrees of simplicity and precision.

From the point of view of the contents of the book, most of the material used to illustrate and address the above computational methods and issues has been drawn from the specialized literature on the reliability and risk analyses of complex systems. The specific contents are limited to a number of relevant topics and techniques which, in spite of not being exhaustive of the very extensive subject of reliability and risk analyses, can form the background material for a senior undergraduate or graduate university course on the subject or as basis for the initiation of young researchers to the field. To this aim, several numerical examples have been provided in support to the theory.

Finally, the realization of the book would have not been possible without the support of several people. In particular, I would like to thank Professors George Apostolakis (Massachusetts Institute of Technology), Marzio Marseguerra (Politecnico di Milano) and Drs. Luca Podofillini (Paul Scherrer Institute) and Andrea Zoia (Politecnico di Milano) for their contributions to the development of the Chapters {7}, {2, 4}, {1, 4, 6}, {3} and the examples therein, respectively. Many thanks are also due to Dr. Giulio Gola (Halden Reactor Project) for the initial translation of the Italian lecture notes at the basis of the material of Chapter 7 (it would have been too 'risky' to leave them as such). My last words of acknowledgments go to Francesco Di Maio who is currently pursuing a PhD at the Politecnico di Milano under my supervision: to him goes my deepest gratitude for the careful, precise work and for the unbreaking passion he has put into the editing of the book (perhaps motivated by the suffering he had to go through when studying the subject on the original lecture notes for my course).

Enrico Zio
Milano, July 2008

References

[1]   Apostolakis G.E., PRA/QRA: *An Historical Perspective*, 2006 Probabilistic/Quantitative Risk Assessment Workshop, 29-30 November 2006, Taiwan.

[2]   Farmer, F.R., *The Growth of Reactor Safety Criteria in the United Kingdom*, Anglo-Spanish Power Symposium, Madrid, 1964.

[3]   Garrick, B.J. and Gekler, W.C., *Reliability Analysis of Nuclear Power Plant Protective Systems*, US Atomic Energy Commission, HN-190, 1967.

[4]   WASH-1400, *Reactor Safety Study*, US Nuclear Regulatory Commission 1975.

[5]   NASA, *Probabilistic Risk Assessment Procedures Guide for NASA Managers and Practitioners*, 2002.

[6]   Aven, T., *Foundations of Risk Analysis*, Wiley, 2003.

[7]   Bedford, T. and Cooke, R., *Probabilistic Risk Analysis*, Cambridge University Press, 2001.

[8]   Henley, E.J. and Kumamoto, H., *Probabilistic Risk Assessment*, NY, IEEE Press, 1992.

[9]   Kaplan, S. and Garrick, B. J., *Risk Analysis*, 1, p. 1-11, 1984.

[10]   McCormick, N.J., *Reliability and Risk Analysis*, New York, Academic Press, 1981.

[11]   NUREG/CR-2300, *PRA Procedures Guide*, Vols. 1&2, January 1983.

[12]   Helton J.C., *Alternative Representations of Epistemic Uncertainty*, Special Issue of Reliability Engineering and System Safety, Vol. 85, 2004.

[13]   Apostolakis G.E., *The Concept of Probability in Safety Assessments of Technological Systems*, Science, 1990, pp. 1359-1364.

[14]   Cai K.-Y., *System Failure Engineering and Fuzzy Methodology. An Introductory Overview*, Fuzzy Sets and Systems 83, 1996, pp. 113-133.

[15]   Da Ruan, Kacprzyk J. and Fedrizzi M. Eds., *Soft Computing for Risk Evaluation and Management*, Physica-Verlag, 2001.

[16]   Soft Methods in Safety and Reliability, Special Sessions I-III, Proceedings of ESREL 2007, Stavanger, Norway, 25-27 June 2007, Volume 1.

[17] Moore R.E., *Methods and Applications of Interval Analysis*, Philapdelphia, PA: SIAM, 1979.

[18] Coolen, F.P.A., *On the Use of Imprecise Probabilities in Reliability*, Quality and Reliability Engineering International, 2004, 20, pp. 193-202.

[19] Coolen, F.P.A. and Utkin, L.V., *Imprecise Probability: A Concise Overview*, In Aven, T. & Vinnem, J.E. (eds) Risk, reliability and societal safety, Proceedings of the European Safety and Reliability Conference (ESREL), Stavanger, Norway, 25-27 June 2007, London, Taylor & Francis.

[20] Utkin, L.V. and Coolen, F.P.A., *Imprecise Reliability: An Introductory Overview*, In Levitin, G. (ed.) Computational Intelligence in Reliability Engineering – New Metaheuristics, Neural and Fuzzy Techniques in Reliability, Springer, 2007.

[21] Zadeh, L.A., *Probability Measures of Fuzzy Events*, Journal of Mathematical Analysis and Applications, 23, 1968, pp. 421-427.

[22] Klir G.J., Yuan B., *Fuzzy Sets and Fuzzy Logic: Theory and Applications*, Prentice Hall, 1995.

[23] Gudder, S., *What is fuzzy probability theory?*, Foundations of Physics 30(10), 2000, pp. 1663-1678.

[24] Zadeh L.A., *Fuzzy Sets*, Information and Control, Vol. 8, 1965, pp. 338-353.

[25] Unwin, S.D., *A fuzzy Set Theoretic Foundation For Vagueness in Uncertainty Analysis*, Risk Analysis 6(1), 1986, pp. 27-34.

[26] Dubois D. and Prade H., *Possibility Theory: An Approach to Computerized Processing of Uncertainty*, New York, Plenum Press, 1988.

[27] Shafer G., *A Mathematical Theory of Evidence*, Princeton, NJ: Princeton University Press, 1976.

[28] Flage, R., Aven, T. and Zio E., *Alternative Representations of Uncertainty in System Risk and Reliability Analysis: Review and Discussion*, Proceedings of ESREL 2008, Valencia Spain, 22-25 September 2008.

1. MARKOV RELIABILITY AND AVAILABILITY ANALYSIS

1.1 Introduction

Let us consider a system whose behaviour is described by its states and the possible transitions among these states. The various system states are defined by the states of the components comprising the system. The components are not restricted to having only two possible states but rather may have a number of different states such as functioning, in standby, degraded, partially failed, completely failed, under maintenance, etc.; the various failure modes of a component may also be defined as states. The transitions between the states occur randomly in time, because caused by various mechanisms and activities such as failures, repairs, replacements and switching operations, which are random in nature. Common cause failures may also be included as possible transitions occurring randomly in time.

Under specified conditions, the stochastic process of the system evolution may be described as a Markov process in which the system states and the possible transitions can be depicted with the aid of a state-space diagram, known as a Markov diagram and be mathematically described by a probabilistic Markov model of equations [1-6].

1.2 Discrete-time, discrete-state Markov processes

1.2.1 The conceptual model

The system can occupy a finite or countably infinite number of states, which can be numbered in a given order 0,1,2,3,4,... . The set of states through which the system can move is the state-space of the random process. The states are mutually exclusive and exhaustive, since the system must be in one state at all times and only in one each time. During the process evolution, the system can move from one state to another, stochastically.

Let us consider an integer random variable which describes the random process of system transition in time, from one state to another. The states occupied by the system at different times are indicated by the values assumed by the random variable in correspondence of such times. The random process may be observed at discrete times or continuously. The first case leads to a discrete-time process of random transitions among discrete system states; the second case leads to a discrete state process, continuous in time. In both cases, the random process of system evolution may be described by a Markov process, discrete or continuous in time, that visits a finite or countably infinite number of states.

We firstly introduce the discrete-time, discrete-state Markov process. The transitions occur at discrete times t_1, t_2, ..., t_n with $t_n = t_{n-1} + \Delta t(n)$. The interval $\Delta t(n)$ between two successive times t_{n-1} and t_n is small such that only one event can occur. For simplicity, we shall assume that the time interval Δt between two successive times t_{n-1}, t_n is always the same independently of n.

The quantification of the system stochastic process evolution amounts to computing the probability that the system is in a given state at a given time, for all possible states and times. To this aim, we need to define the rules that govern the system transitions and assign them appropriate values of probability of occurrence.

Let us denote by $X(n)$ the random variable indicating the state of the system at time t_n. This random variable gives an information about the system state in correspondence of the observation time t_n. For example, $X(5) = 3$ indicates that at the fifth time step the system is in state 3.

In general, the probability of a future state of the system may depend on its entire life history. Thus,

$$P[X(n+1) = j \mid X(0) = x_0, X(1) = x_1, X(2) = x_2, \ldots\ldots, X(n) = x_n] \quad (1.1)$$

The fundamental assumption characterizing a Markov process is that the future state of the system depends solely on its present state, thus

$$P[X(n+1) = x_j \mid X(n) = x_n] \quad (1.2)$$

Since the probability of the system moving to a next state is independent of its past history, the random process of system evolution is said to have no memory: all that is relevant is the present state of the system and not the history to get there.

Considering two arbitrary times t_m, t_n ($t_n > t_m$), we introduce the transition probability that the system in state i at time t_m moves to state j at time t_n (Fig. 1.1):

$$p_{ij}(m,n) = P[X(n) = j \mid X(m) = i] \quad n > m \quad (1.3)$$

Fig. 1.1

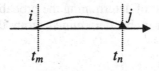

If $p_{ij}(m,n)$ depends only on the interval $t_n - t_m$, and not on the individual times t_m and t_n, the Markov process is said to be homogeneous in time and the transition probabilities are said to be stationary. For a generic time interval $k \cdot \Delta t$, we then have (Fig. 1.2)

$$p_{ij}(k) = P[X(k) = j \mid X(0) = i] = P[X(k+s) = j \mid X(s) = i] \quad s \geq 0 \quad (1.4)$$

Fig. 1.2

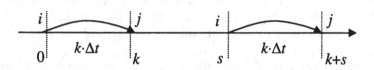

The transition probabilities satisfy the following properties:

$$p_{ij}(k) \geq 0 \text{ for } k > 0 \tag{1.5}$$

$$\sum_{\text{all } j} p_{ij}(k) = 1 \text{ for } k > 0 \tag{1.6}$$

$$p_{ij}(n = k + r) = \sum_{\text{all } l} p_{il}(k) \cdot p_{lj}(r) \text{ for } k, r > 0 \tag{1.7}$$

The last property is represented pictorially in Fig. 1.3 below and follows from the Markov assumption and the theorem of total probability:

Fig. 1.3

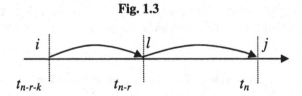

The problem is then that of determining the probability of transition at the k-th time step. This can be determined from the one step transition probabilities:

$$p_{ij}(1) = P[X(n+1) = j \mid X(n) = i] = p_{ij} \tag{1.8}$$

The p_{ij}'s are the one step transition probabilities of an homogeneous Markov process dependent only on the length of the time interval k, which is not written explicitly. Considering a finite state-space with $N+1$ states, we need to provide all one step transition probabilities from any state i to any other state j, $j = 0,1,2,...,$ N. These probabilities can be arranged in a $((N+1)\times(N+1))$ transition probability matrix, $\underline{\underline{A}}$,

$$
\underline{\underline{A}} = \begin{matrix} & \begin{matrix} i/j & \quad 0 & \quad 1 & \cdots & N \end{matrix} \\ \begin{matrix} 0 \\ 1 \\ \cdots \\ N \end{matrix} & \begin{pmatrix} p_{00} & p_{01} & \cdots & p_{0N} \\ p_{10} & p_{11} & \cdots & p_{1N} \\ \cdots & \cdots & \cdots & \cdots \\ p_{N0} & p_{N1} & \cdots & p_{NN} \end{pmatrix} \end{matrix} \tag{1.9}
$$

The transition probability matrix has the following properties:

1. $0 \le p_{ij} \le 1$ $\quad \forall\, i,j \in \{0,1,2,...,N\}$, since all matrix elements are probabilities

2. $\sum_{j=0}^{N} p_{ij} = 1$ $\quad i = 0,1,2,...,N$, since the states are assumed exhaustive

With these properties, the matrix $\underline{\underline{A}}$ is said to be a stochastic matrix. Moreover, given property 2 only $(N+1)\times N$ elements need to be provided for $\underline{\underline{A}}$ to be fully known. The matrix $\underline{\underline{A}}$ contains the fundamental data and, together with the system states and transitions, describes the stochastic process of system evolution.

1.2.2 State probabilities

Given an N-state Markov process, we introduce the row vector of the probabilities of the system being in state 1, 2, ... , N at the n-th time step

$$
\underline{P}(n) = \begin{bmatrix} P_0(n) & P_1(n) & \cdots & P_N(n) \end{bmatrix} \tag{1.10}
$$

At the time step $n=0$, this vector is initialized to:

$$
\underline{P}(0) = \underline{C} = \begin{bmatrix} C_0 & C_1 & \cdots & C_N \end{bmatrix} \tag{1.11}
$$

The vector $\underline{P}(n)$ at the n-th time step can be induced in terms of those at the previous time steps by repeatedly applying the theorem of total probability. At the first time step, $n=1$, we have:

$$P_j(1) = P[X(1) = j]$$
$$= \sum_{i=0}^{N} P[X(1) = j | X(0) = i] \cdot P[X(0) = i]$$
$$= \sum_{i=0}^{N} p_{ij} C_i = p_{0j} \cdot C_0 + p_{1j} \cdot C_1 + p_{2j} \cdot C_2 + \ldots + p_{Nj} \cdot C_N,$$
$$\text{with } j = 0,1,2,\ldots,N$$

(1.12)

which in matrix notation reads:

$$\underline{P}(1) = \underline{C} \cdot \underline{\underline{A}}$$

(1.13)

At the successive step, $n=2$, we get:

$$P_j(2) = P[X(2) = j]$$
$$= \sum_{k=0}^{N} P[X(2) = j | X(1) = k] \cdot P[X(1) = k]$$
$$= \sum_{k=0}^{N} p_{kj} \cdot P_k(1)$$
$$= P_0(1) \cdot p_{0j} + P_1(1) \cdot p_{1j} + P_2(1) \cdot p_{2j} + \ldots + P_N(1) \cdot p_{Nj},$$
$$\text{with } j = 0,1,2,\ldots,N$$

(1.14)

which in matrix form becomes:

$$\underline{P}(2) = \underline{P}(1) \cdot \underline{\underline{A}} = (\underline{C}\,\underline{\underline{A}})\underline{\underline{A}} = \underline{C}\,\underline{\underline{A}}^2$$

(1.15)

Proceeding in the same way, at the n-th step we get:

$$\underline{P}(n) = \underline{C} \cdot \underline{\underline{A}}^n$$

(1.16)

which in matrix form represents the fundamental equation describing in a comprehensive way the random transition process in the state-space.

1.2.3 Multi-step transition probabilities

The n-th power of matrix $\underline{\underline{A}}$ represents the n-step transition probability matrix:

$$\underline{\underline{A}}^n = \begin{pmatrix} p_{00}(n) & p_{01}(n) & \cdots & p_{0N}(n) \\ p_{10}(n) & p_{11}(n) & \cdots & p_{1N}(n) \\ \cdots & \cdots & \cdots & \cdots \\ p_{N0}(n) & p_{N1}(n) & \cdots & p_{NN}(n) \end{pmatrix} \tag{1.17}$$

whose generic element $p_{ij}(n)$ is the probability of arriving in state j after n steps, given that the initial state was i, i.e.:

$$p_{ij}(n) = P\left[X(n) = j \mid X(0) = i\right] \tag{1.18}$$

Note that $p_{ij}(n)$ is the sum of the probabilities of all trajectories with length n which originate in state i and end in state j (see the diagram in Fig. 1.4 for transitions between states 2 and 3).

Fig. 1.4

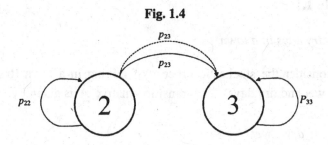

$$p_{23}(2) = p_{22} \cdot p_{23} + p_{23} \cdot p_{33}$$

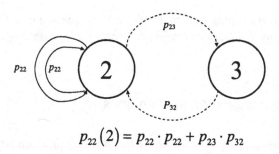

$$p_{22}(2) = p_{22} \cdot p_{22} + p_{23} \cdot p_{32}$$

An alternative approach to the evaluation of the $p_{ij}(n)$ consists in using the Chapman-Kolmogorov equation (Fig. 1.5):

$$p_{ij}(n) = P\left[X(n) = j \mid X(0) = i\right]$$
$$= \sum_{l} P\left[X(n) = j \mid X(r) = l\right] \cdot P\left[X(r) = l \mid X(0) = i\right] \tag{1.19}$$

Fig. 1.5

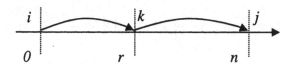

Example 1.1

Wet and dry days in a town.

Let us consider the stochastic process of raining in a town (transitions between wet and dry days). The transition matrix $\underline{\underline{A}}$ is given by:

$$\underline{\underline{A}} = \begin{array}{c} \\ dry \\ wet \end{array} \begin{array}{cc} dry & wet \\ \begin{pmatrix} 0.8 & 0.2 \\ 0.5 & 0.5 \end{pmatrix} \end{array}$$

For example, the element (2, 2) represents the conditional probability that if it is wet today it will also be wet tomorrow. The process may be illustrated by the Markov diagram in Fig. 1.6 where state 1 indicates a dry day and state 2 a wet one:

Fig. 1.6

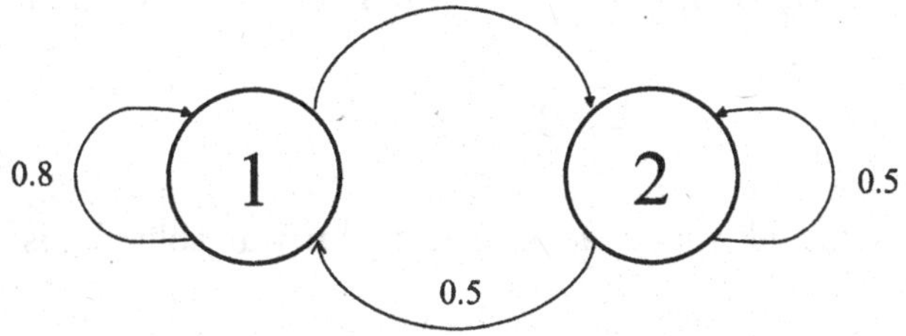

Question: If today the weather is dry, what is the probability that it will be dry two days from now?

Answer: Starting from the initial condition $\underline{C} = \begin{bmatrix} 1 & 0 \end{bmatrix}$, at $n=2$ time steps we get:

$$\underline{P}(2) = \begin{bmatrix} 1 & 0 \end{bmatrix} \cdot \begin{bmatrix} 0.8 & 0.2 \\ 0.5 & 0.5 \end{bmatrix} \cdot \begin{bmatrix} 0.8 & 0.2 \\ 0.5 & 0.5 \end{bmatrix} = \begin{bmatrix} 0.74 & 0.26 \end{bmatrix}$$

The first value in the vector, 0.74, represents the probability of the conditions being dry two days from now.

1.2.4 Solution of the fundamental equation

Let us now return to the problem of solving the fundamental equation of the Markov process:

$$\begin{cases} \underline{P}(n) = \underline{C}\,\underline{A}^n \\ \underline{P}(0) = \underline{C} \end{cases} \tag{1.20}$$

This system of equations can be solved using the eigenvalue method. First, we solve the associated eigenvalue problem:

$$\underline{V} \cdot \underline{\underline{A}} = \omega \cdot \underline{V} \qquad (1.21)$$

where ω is a scalar and $\underline{V}$ is an eigenvector.

The above equations may be written in an homogeneous form as:

$$\underline{V} \cdot \left(\underline{\underline{A}} - \omega \cdot \underline{\underline{I}} \right) = 0 \qquad (1.22)$$

where $\underline{\underline{I}}$ is the identity matrix. The non-trivial solution is found by setting:

$$\det \left(\underline{\underline{A}} - \omega \cdot \underline{\underline{I}} \right) = 0 \qquad (1.23)$$

from which we get the eigenvalues ω_j, $j = 0, 1, ..., N$. Substituting these values back into (1.21):

$$V_j \cdot \underline{\underline{A}} = \omega_j \cdot \underline{V}_j \qquad (1.24)$$

we get the $N+1$ corresponding eigenvectors $\underline{V}_j$, $j = 0, 1, ..., N$.

The eigenvectors $\underline{V}_j$ span the $N+1$-dimensional space and can be used as basis to write any vector as a linear combination of them. Hence, the unknown probability vector $\underline{P}(n)$, after n steps of time, can be written as:

$$\underline{P}(n) = \sum_{j=0}^{N} \alpha_j \cdot \underline{V}_j \qquad (1.25)$$

and similarly the known initial condition vector can be written as:

$$\underline{C} = \sum_{j=0}^{N} c_j \cdot \underline{V}_j \qquad (1.26)$$

The problem is then reduced to determining the expansion coefficients α_j and c_j.

To determine the values for c_j, we resort to the adjoint eigenvalue problem, recalling that the adjoint of a real matrix is simply its transpose:

$$\underline{V}_j^+ \cdot \underline{\underline{A}}^T = \omega_j^+ \cdot \underline{V}_j^+ \,, \quad j = 0,1,...,N \qquad (1.27)$$

The adjoint eigenvalues ω_j^+ depend only on the determinant of the transpose matrix $\underline{\underline{A}}^T$, which is equal to the determinant of the original matrix $\underline{\underline{A}}$. Thus, the adjoint eigenvalues ω_j^+ are the same as ω_j. However the adjoint eigenvectors $\underline{V}_j^+$ are different from $\underline{V}_j$. By definition of the adjoint problem and taking into account that $\underline{V}_j^+$ and $\underline{V}_j$ are orthonormal vectors, we have:

$$<\underline{V}_j^+, \underline{V}_i> \equiv \underline{V}_j^+ \cdot \underline{V}_i^T = \begin{cases} 0 \text{ if } i \neq j \\ k \text{ otherwise} \end{cases} \qquad (1.28)$$

where k is a real value.

Multiplying the left-hand side of $\underline{C} = \sum_{i=0}^{N} c_i \underline{V}_i$ by the adjoint eigenvector $\underline{V}_j^+$, we get:

$$<\underline{V}_j^+, \underline{C}> = \sum_{i=0}^{N} c_i <\underline{V}_j^+, \underline{V}_i> = c_j <\underline{V}_j^+, \underline{V}_i> \to c_j = \frac{<\underline{V}_j^+, \underline{C}>}{<\underline{V}_j^+, \underline{V}_i>} \qquad (1.29)$$

To determine the values of α_j we have:

$$\underline{P}(n) = \sum_{j=0}^{N} \alpha_j \cdot \underline{V_j}$$

$$\underline{C} = \sum_{j=0}^{N} c_j \cdot \underline{V_j} \qquad (1.30)$$

$$\underline{P}(n) = \underline{C}\underline{\underline{A}}^n$$

Substituting the second equation into the third one and setting the resulting equation equal to the first one we get:

$$\sum_{j=0}^{N} \alpha_j \cdot \underline{V_j} = \left(\sum_{j=0}^{N} c_j \cdot \underline{V_j} \right) \cdot \underline{\underline{A}}^n \qquad (1.31)$$

Also, from:

$$\underline{V_j}\underline{\underline{A}} = \omega_j \cdot \underline{V_j} \qquad (1.32)$$

we have:

$$\underline{V_j}\underline{\underline{A}}^n = \omega_j^n \cdot \underline{V_j} \qquad (1.33)$$

and substituting into the previous equation:

$$\sum_{j=0}^{N} \alpha_j \cdot \underline{V_j} = \sum_{j=0}^{N} c_j \cdot \omega_j^n \cdot \underline{V_j} \qquad (1.34)$$

which yields:

$$\alpha_j = c_j \cdot \omega_j^n \qquad (1.35)$$

Knowing the $\underline{V_j}$ and α_j, $j = 0,1,...,N$, the probability vector $\underline{P}(n)$ in (1.25) is completely determined.

Example 1.2

Consider the occupancy history of a 'slot' in a transit system. A continuous moving 'belt' of 'slots' (each slot designed for a single passenger) passes along the transit route. The route has a large number of stations, in which each individual slot slows down. If it is full, its passenger may exit with probability 0.8; if it is empty, a waiting user, if any present, will fill it. The probability that a person will be present at the station is 0.5. Assume that the slot starts empty.

Consider the process $X(n)$ defined by:

$X(n) = 0$ if the slot is empty between the n-th station and the $(n+1)$-th station

$X(n) = 1$ if the slot is full between the n-th station and the $(n+1)$-th station.

a) Show that $X(n)$ is a Markov process
b) Find the transition probability matrix
c) Find the probability that the slot will be empty as it leaves the n-th station.
d) Find the stationary distribution. How long does it take to reach it?
e) What is the probability $P\{E_n\}$ of the event $E_n=\{$the slot is full for the first time at station $n\}$?

Solution

a) From the problem description, it can be inferred that the transition probabilities depend only on the current state of the system. Therefore, it is a Markov process, discrete in time (in this case 'in stations') and states.

b) The transition probability matrix $\underline{A}$ governs the transitions from a generic station n to the successive station $n+1$. Let p_{ij} denote the elements of the transition probability matrix $\underline{A}$, $i=0,1,2,...,N$, $j=0,1,2,...,N$. In our case $N=1$ since there are two possible system states ($X(n)=0$ and $X(n)=1$).

The first element p_{00} of the matrix $\underline{\underline{A}}$ is the probability that the slot will be empty at the station $n+1$ ($X(n+1)=0$), given that it is empty at the station n ($X(n)=0$):

$$p_{00} = P\{X(n+1)=0|X(n)=0\} = 0.5$$

In fact, if the slot has left the station n being empty, the probability that it will leave the station $n+1$ still empty is equal to the probability that nobody is waiting at the station $n+1$, namely 0.5.

The second element, p_{01}, is the probability that the slot will be full at the station $n+1$ ($X(n+1)=1$), given that it is empty at the station n ($X(n) = 0$).

$$p_{01} = P\{X(n+1)=1|X(n)=0\} = 0.5$$

Evidently, the slot will be filled with a probability equal to that of somebody waiting at station $n+1$, namely 0.5.

The probability that a slot will be empty at the station $n+1$ ($X(n+1)=0$), given that it is full at the station n ($X(n)=1$) is the probability that a passenger gets out of the slot times the probability that somebody is not waiting at the $(n+1)$-th station:

$$p_{10} = P\{X(n+1)=0|X(n)=1\} = 0.8 \cdot 0.5 = 0.4$$

Finally, if a slot is full at the station n, it will be full at station $n+1$ either if the passenger does not get out of it or if the passenger gets out and another fills its place:

$$p_{11} = P\{X(n+1)=1|X(n)=1\} = 0.2 + 0.8 \cdot 0.5 = 0.6 .$$

Thus, the transition matrix $\underline{\underline{A}}$ reads:

$$\underline{\underline{A}} = \begin{bmatrix} 0.5 & 0.5 \\ 0.4 & 0.6 \end{bmatrix}$$

c) Let $\underline{P}(n)$ be the row vector of state probabilities:

$$\underline{P}(n) = [P_0(n) \quad P_1(n)]$$

where $P_i(n)$ is the probability of being in state i at the n-th station.
From the theory, $\underline{P}(n)$ can be expressed as a linear combination of the eigenvectors of $\underline{\underline{A}}$, $\underline{V}_j$, $j=0,1,...,N$. The eigenvalues and eigenvectors, ω_j and $\underline{V}_j$, $j=0,1,...,N$, respectively, are found as usual by solving the linear system of equations:

$$\underline{V} \cdot (\underline{\underline{A}} - \omega \underline{\underline{I}}) = 0$$

Then, $\underline{P}(n)$ can be written as:

$$\underline{P}(n) = \sum_{j=0}^{N} c_j \cdot \omega_j^n \cdot \underline{V}_j$$

where from the adjoint eigenvalue/eigenvector problem (1.27) and the orthonormality property (1.28):

$$c_j = \frac{<\underline{V}_j^+, \underline{C}>}{<\underline{V}_j^+, \underline{V}_j>}$$

with the following notation:

$\underline{C}$ = vector denoting the initial conditions. In our case, assuming that the slot starts empty, $\underline{C} = [1 \ 0]$.

$\underline{V}_j^+$ = j-th eigenvector of the adjoint problem, $j=0,1$, satisfying the system (1.27) with $\omega_j^+ = \omega_j$.

Let us first find the direct eigenvalues and eigenvectors, ω_j and $\underline{V_j}$, j=0,1. Solving the homogeneous system $\underline{V} \cdot (\underline{\underline{A}} - \omega \underline{\underline{I}}) = 0$:

$$\det\left[\underline{\underline{A}} - \omega \underline{\underline{I}}\right] = \det\begin{bmatrix} 0.5 - \omega & 0.5 \\ 0.4 & 0.6 - \omega \end{bmatrix}$$

$$= (0.5 - \omega) \cdot (0.6 - \omega) - 0.2 = \omega^2 - 1.1\omega + 0.1 = 0$$

The roots which render the determinant equal to zero are:

$$\Rightarrow \quad \omega_0 = 1$$
$$\omega_1 = 0.1$$

and the first eigenvector $\underline{V_0} \equiv \begin{bmatrix} V_0^0 & V_0^1 \end{bmatrix}$ is found replacing ω_0 in $\underline{V} \cdot (\underline{\underline{A}} - \omega \underline{\underline{I}}) = 0$:

$$\underline{V_0}\begin{bmatrix} 0.5 & 0.5 \\ 0.4 & 0.6 \end{bmatrix} - \underline{V_0} = 0$$

$$\begin{cases} 0.5V_0^0 + 0.4V_0^1 - V_0^0 = 0 \\ 0.5V_0^0 + 0.6V_0^1 - V_0^1 = 0 \end{cases}$$

$$\Rightarrow \qquad 0.5V_0^0 - 0.4V_0^1 = 0 \Rightarrow \qquad \underline{V_0} = \begin{bmatrix} 0.8 & 1 \end{bmatrix} V_0^1$$

Similarly, for the second eigenvector $\underline{V_1} \equiv \begin{bmatrix} V_1^0 & V_1^1 \end{bmatrix}$, replacing ω_1 in $\underline{V} \cdot (\underline{\underline{A}} - \omega \underline{\underline{I}}) = 0$:

$$\underline{V_1}\begin{bmatrix} 0.5 & 0.5 \\ 0.4 & 0.6 \end{bmatrix} - 0.1\underline{V_1} = 0.$$

Considering the first equation of the above homogeneous system (which, as usual, is undetermined):

$$0.5V_1^0 + 0.4V_1^1 - 0.1V_1^0 = 0$$

$$\Rightarrow \qquad 0.4V_1^0 + 0.4V_1^1 = 0 \qquad \Rightarrow \qquad \underline{V}_1 = [-1 \ +1] \ V_1^1$$

Analogously, the adjoint eigenvectors are:

$$\underline{V}_0^+ \begin{bmatrix} 0.5 & 0.4 \\ 0.5 & 0.6 \end{bmatrix} - \underline{V}_0^+ = 0$$

$$0.5(V_0^0)^+ + 0.5(V_0^1)^+ - (V_0^0)^+ = -(0.5V_0^0)^+ + (0.5V_0^1)^+ = 0$$

$$\Rightarrow \qquad \underline{V}_0^+ = [+1 \ +1] \ (V_1^1)^+$$

$$\underline{V}_1^+ \begin{bmatrix} 0.5 & 0.4 \\ 0.5 & 0.6 \end{bmatrix} - 0.1V_1^+ = 0$$

$$0.5(V_1^0)^+ + 0.5(V_1^1)^+ - 0.1(V_1^0)^+ = 0.4(V_1^0)^+ + 0.5(V_1^1)^+$$

$$\Rightarrow \qquad \underline{V}_1^+ = [1 \ -0.8] \ (V_1^1)^+ \ .$$

We now have all the quantities used in (1.29) to calculate the coefficients c_j of the probability vector expansion $\underline{P}(n) = \sum_{j=0}^{N} c_j \cdot \omega_j^n \underline{V}_j$:

$$c_0 = \frac{<\underline{V}_0^+, \underline{C}>}{<\underline{V}_0^+, \underline{V}_0>} = \frac{[1 \ 1] \cdot [1 \ 0]^T}{[1 \ 1] \cdot [0.8 \ 1]^T} = \frac{1}{1.8}$$

$$c_1 = \frac{<\underline{V}_1^+, \underline{C}>}{<\underline{V}_1^+, \underline{V}_1>} = \frac{[1 \ -0.8] \cdot [1 \ 0]^T}{[1 \ -0.8] \cdot [-1 \ 1]^T} = -\frac{1}{1.8}$$

The vector of the state probabilities $\underline{P}(n)$ is thus:

$$\underline{P}(n) = c_0 \alpha_0^n \underline{V}_0 + c_1 \alpha_1^n \underline{V}_1 = \frac{1}{1.8}[0.8 \quad 1] - \frac{1}{1.8}[-1 \quad +1] \cdot (0.1)^n$$

$$= \left[\frac{0.8}{1.8} + \frac{0.1^n}{1.8} \quad \frac{1}{1.8} - \frac{0.1^n}{1.8} \right]$$

The probability that the slot will be empty at the n-th station, $P_0(n)$ is thus:

$$P_0(n) = 0.444 + 0.56 \cdot 0.1^n$$

d) The stationary value for the probability that the slot will be empty at a station can be obtained by taking the limit of $P_0(n)$ by letting $n \to \infty$:

$$P_0(\infty) = \lim_{n \to \infty} P_0(n) = 0.444$$

Proceeding step-by-step to find after how many stations the asymptotic solution is practically attained:

$P_0(0) = 1$
$P_0(1) = 0.5$
$P_0(2) = 0.45$
$P_0(3) = 0.445$
$P_0(4) = 0.4445$

So, the asymptotic solution is practically reached in four steps.

e) The event $E_n = \{$the slot is full for the first time at station $n\}$ requires that the slot is in the empty state for $n - 1$ consecutive stations and that it transfers to the full state at the n-th station. Then, it is given by the geometric distribution with $p_{00} = 0.5$.

$$\Pr\{E_n\} = \begin{cases} 0 & n = 0 \\ (p_{00})^{n-1} p_{10} = (0.5)(0.5)^{n-1} = (0.5)^n & n > 0 \end{cases}$$

1.2.5   Steady state probabilities for ergodic systems

We recall that an ergodic set of states is one in which all states communicate and which cannot be left once it is entered [6]. In other words, it is a collective absorbing state also called "chain". Note that all finite Markov processes must have at least one chain, because a finite Markov process cannot have all transient states since the process would keep leaving its current state to go somewhere else.

For ergodic systems, we can obtain the steady state probabilities $\Pi_j, j = 0,1,2,...,N$ of the system being in state j asymptotically. Considering that the eigenvalue of the fundamental mode is $\omega_0 = 1$ whereas all the others are $|\omega_j| < 1$, $j = 1,2,...,N$, at steady state we have [6]:

$$\lim_{n \to \infty} \underline{P}(n) = \lim_{n \to \infty} \sum_{j=0}^{N} \alpha_j \cdot \underline{V}_j = \lim_{n \to \infty} \sum_{j=0}^{N} c_j \cdot \omega_j^n \cdot \underline{V}_j = c_0 \underline{V}_0 = \underline{\Pi} \qquad (1.36)$$

More simply, the steady state solution can be found from the recursive equation:

$$\underline{P}(n) = \underline{P}(n-1) \cdot \underline{\underline{A}} \qquad (1.37)$$

and considering that at steady state

$$\underline{P}(n) = \underline{P}(n-1) = \underline{\Pi} \qquad (1.38)$$

we obtain:

$$\underline{\Pi} = \underline{\Pi} \cdot \underline{\underline{A}} \qquad (1.39)$$

which can be solved accounting for the normalization of the probabilities on the mutually exclusive and exhaustive states, $\sum_{j=0}^{N} \Pi_j = 1$.

Example 1.3

Wet and dry days in a town.

Returning to the Example 1.1,

Given:

$$\begin{array}{cc} & dry \quad wet \\ \underline{A} = \begin{array}{c} dry \\ wet \end{array} & \begin{pmatrix} 0.8 & 0.2 \\ 0.5 & 0.5 \end{pmatrix} ; \quad \underline{C} = \begin{bmatrix} 1 & 0 \end{bmatrix} \end{array}$$

Question: What is the probability that one year from now the day will be dry?

Answer: $\underline{P}(365\,days = 1\,year) = \underline{C}\underline{P}^{365}$

The evaluation of $\underline{P}^{365}$ requires a lot of computations. On the other hand, we can reasonably assume that at $n=365$ the steady state condition is established so that:

$$\begin{cases} \Pi_1 = 0.8 \cdot \Pi_1 + 0.5 \cdot \Pi_2 \\ \Pi_1 + \Pi_2 = 1 \end{cases} \Rightarrow \underline{\Pi} = \begin{bmatrix} 0.714 & 0.286 \end{bmatrix}$$

The answer to the question is then $\Pi_1 = 0.714$.

1.2.6   First passage probabilities

We introduce the so called "first passage" probability:

$$f_{ij}(n) = P\left[X(n) = j \ \text{for the first time} \middle| X(0) = i \right] \qquad (1.40)$$

which represents the probability to arrive for the first time in state j after n steps, having departed state i at the initial time $n=0$.

For simplicity let us consider a two state Markov chain as in Fig. 1.7:

Fig. 1.7

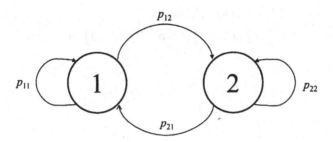

Then,

$f_{11}(1) = p_{11}$ is the probability of going from state 1 to state 1 in 1 step for the first time.

$f_{11}(n) = p_{12} \cdot p_{22}^{n-2} \cdot p_{21}$ is the probability that the system starting from state 1 will return to the same state 1 for the first time after n steps; in other words, it is the probability to depart from state 1 and then jump back at the n-th step to the initial state 1: this is achieved by jumping in state 2 at the first step (p_{12}), remaining in state 2 during the successive $n-2$ steps (p_{22}^{n-2}) and moving back in the initial state 1 at the n-th step (p_{21}).

$f_{12}(n) = p_{11}^{n-1} \cdot p_{12}$ is the probability that the system will arrive for the first time in state 2 after n steps; this is equal to the probability of remaining in state 1 for $n-1$ steps (p_{11}^{n-1}) and then jumping in state 2, at the final step (p_{12}).

In order to determine the first passage probabilities $f_{ij}(n)$, $n = 1, 2, 3, \ldots$ we can adopt an iterative procedure as follows:

$$f_{ij}(1) = p_{ij}(1) = p_{ij}$$
$$f_{ij}(2) = p_{ij}(2) - f_{ij}(1) \cdot p_{jj}$$
$$f_{ij}(3) = p_{ij}(3) - f_{ij}(1) \cdot p_{jj}(2) - f_{ij}(2) \cdot p_{jj}$$

...   (1.41)

$$f_{ij}(k) = p_{ij}(k) - \sum_{l=1}^{k-1} f_{ij}(k-l) p_{jj}(l)$$

...

For example, the second term in $f_{ij}(2)$, $f_{ij}(1) \cdot p_{jj}$ is the probability of going for the first time from state i to state j after the first step and then remaining there at the successive step: to compute the probability $f_{ij}(2)$ of arriving from state i to state j exactly at the second step, the probability of this sequence of transitions must be taken out of the total probability of finding the system in state j after two steps, starting from state i, i.e. $p_{ij}(2)$.

Now we are in the position to compute the probability $q_{ij}(m)$ that the system goes from an initial state i to another state j within m steps, as sum of the probabilities of the mutually exclusive events of reaching j for the first time after $n=1,2,3,\ldots,m$ steps:

$$q_{ij}(m) = \sum_{n=1}^{m} f_{ij}(n)$$ (1.42)

The probability $q_{ij}(\infty)$ of eventually reaching state j from state i is then:

$$q_{ij}(\infty) = \lim_{m \to \infty} q_{ij}(m)$$ (1.43)

Denoting by f_{ii} the probability of eventually returning to the initial state:

$$f_{ii} = q_{ii}(\infty)$$ (1.44)

A state i is said to be "recurrent" if the system starting at such state will surely (probability equal to 1) return to it sooner or later, i.e. [6]:

$$f_{ii} = q_{ii}(\infty) = 1 \qquad (1.45)$$

For recurrent states there exists a steady state probability, $\Pi_i \neq 0$.

On the contrary, state i is a "transient" state if the system in state i has a finite probability of never returning to it, i.e.:

$$f_{ii} = q_{ii}(\infty) < 1 \qquad (1.46)$$

For these states, at steady state $\Pi_i = 0$. Thus, the steady state probability will be non-zero only for the recurrent states. Obviously, we cannot have a finite Markov process in which all states are transients because eventually it will leave them and somewhere it must go at steady state.

We also define "absorbing" states those for which once the system enters it can never leave, i.e.:

$$p_{ii} = 1 \qquad (1.47)$$

If an absorbing state exists in a Markov chain, the system will eventually reach it and be trapped there.

Finally, another quantity of interest is the average occupation time of state i, l_i, which represents the number of steps before the system exits that state:

$$l_i = \frac{1}{1 - p_{ii}} \qquad (1.48)$$

1.3 Continuous time, discrete-state Markov processes

1.3.1 The conceptual model

Let us consider a system which may stay in $N+1$ configurations, $j = 0,1,2,...,N$. The state variable describing the system configuration at time t is denoted by $X(t)$. The system is assumed to start in a specified state, say i, at time $t=0$. The transitions between states are assumed to occur continuously in time as described by a stochastic process $\{X(t); t \geq 0\}$ governed by the transition probabilities.

For many systems, the transitions are well described by a stochastic process with the Markov property: given that a system is in state i at time t [i.e., $X(t)=i$], the probability of reaching state j at time $t+v$ does not depend on the states $X(u)$ visited by the system prior to t ($0 \leq u < t$). In other words, given the present state $X(t)$ of the system, its future behavior is independent of the past:

$$P\left[X(t+v) = j \middle| X(t) = i, X(u) = x(u), 0 \leq u < t \right]$$
$$= P\left[X(t+v) = j \middle| X(t) = i \right] \tag{1.49}$$

As illustrated in Section 1.2, the conditional probabilities

$$P\left[X(t+v) = j \middle| X(t) = i \right] \qquad i,j = 0,1,2,3,...,N \tag{1.50}$$

are called the transition probabilities of the Markov process. If the transition probabilities do not depend on time t but only on the time interval v for the transition, then the Markov process is said to be homogeneous or stationary:

$$P\left[X(t+v) = j \middle| X(t) = i \right] = p_{ij}(v) \text{ for } t,v > 0 \text{ and } i,j = 0,1,2,...,N \tag{1.51}$$

A Markov process with stationary transition probabilities has no memory.

Starting from the discrete-time formulation of Markov processes described in the previous Section 1.2 and considering a time step dt sufficiently small that only one event can occur, we write for the one step transition probability from state i to state j:

$$p_{ij}(dt) = P\left[X(t+dt) = j \middle| X(t) = i\right] = \alpha_{ij} \cdot dt + \theta(dt) \tag{1.52}$$

where $\displaystyle\lim_{dt \to 0} \frac{\theta(dt)}{dt} = 0$.

The parameter α_{ij} is the transition rate from state i to state j. Since α_{ij} is constant, the time T_{ij} that the system stays in state i before making a transition to state j is exponentially distributed with parameter α_{ij}.

As in the discrete-time case, we can define a transition probability matrix with the form:

$$\underline{\underline{A}} = \begin{pmatrix} 1 - dt \cdot \displaystyle\sum_{j=1}^{N} \alpha_{0j} & \alpha_{01} \cdot dt & \cdots & \alpha_{0N} \cdot dt \\[2ex] \alpha_{10} \cdot dt & 1 - dt \cdot \displaystyle\sum_{\substack{j=0 \\ j \neq 1}}^{N} \alpha_{1j} & \cdots & \alpha_{1N} \cdot dt \\[2ex] \cdots & \cdots & \cdots & \cdots \end{pmatrix} \tag{1.53}$$

In analogy to the discrete-time case, we write the following fundamental matrix equation which governs the Markov process continuous in time:

$$\underline{P}(t+dt) = \underline{P}(t) \cdot \underline{\underline{A}} \tag{1.54}$$

where, for example, the first equation has the form:

$$P_0(t+dt) = \left[1 - dt \sum_{j=1}^{N} \alpha_{0j}\right] P_0(t) + \alpha_{10} P_1(t) \cdot dt + \ldots + \alpha_{N0} P_N(t) dt \tag{1.55}$$

Subtracting $P_0(t)$ on both sides, dividing by dt and in the limit of $dt \to 0$, we get:

$$\frac{dP_0}{dt} = -\sum_{j=1}^{N} \alpha_{0j} \cdot P_0(t) + \alpha_{10} \cdot P_1(t) + \ldots + \alpha_{N0} \cdot P_N(t) \qquad (1.56)$$

Manipulating in the same way the other equations of the system (1.54), we can write in matrix form:

$$\frac{dP}{dt} = \underline{P}(t) \cdot \underline{\underline{A}}^*, \quad \underline{\underline{A}}^* = \begin{pmatrix} -\sum_{j=1}^{N}\alpha_{0j} & \alpha_{01} & \cdots & \alpha_{0N} \\ \alpha_{10} & -\sum_{\substack{j=0 \\ j\neq 1}}^{N}\alpha_{1j} & \cdots & \alpha_{1N} \\ \cdots & \cdots & \cdots & \cdots \end{pmatrix} \qquad (1.57)$$

The above is a system of linear, first-order differential equations in the unknown state probabilities $P_j(t), j = 0,1,2,\ldots,N, t \geq 0$. The matrix $\underline{\underline{A}}^*$ contains the transition rates of the system; to simplify the notation, from now on the transition rate matrix $\underline{\underline{A}}^*$ will be simply denoted as $\underline{\underline{A}}$. Note that $\alpha_{ii} = \sum_{\substack{j=0 \\ j\neq i}}^{N} \alpha_{ij}$.

Example 1.4

The Poisson process is an infinite Markov chain ($N = \infty$).

The random variable of interest is the number of events observed in a period of time. Thus, the possible system states are $0,1,2,\ldots,\infty$.

Only one event can occur in each small Δt, with probability $\lambda \Delta t$. The probability that the event does not occur in Δt is $1-\lambda \Delta t$, which represents the self-state transition probability, i.e. the probability of remaining in the initial state, which is equal to one minus the sum of the probabilities of leaving that state.

The corresponding Markov diagram is shown in Fig. 1.8:

Fig. 1.8

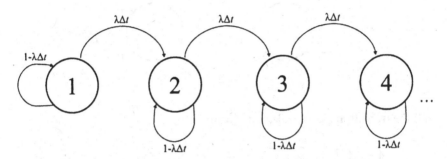

The transition matrix of the Poisson process has infinite dimension:

$$\underline{\underline{A}} = \begin{pmatrix} -\lambda & \lambda & 0 & 0 & ... & 0 \\ 0 & -\lambda & \lambda & 0 & ... & 0 \\ ... & ... & ... & ... & ... & ... \end{pmatrix}$$

Note that the system does not comprise any recurrent state because the chain is infinite (whereas there are always recurrent states in finite Markov chains because the system leaves a state with a finite probability and has to go into some other state).

Example 1.5

One component/one repairman with exponential failure and repair times distributions.

Consider a component which can be in two states, working (0) and failed (1). Let λ and μ be the rates of failure (transitions from state 0 to state 1) and repair (transitions from state 1 to state 0), respectively. The Markov diagram is then:

Fig. 1.9

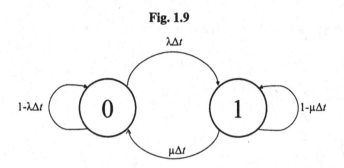

and the transition matrix takes the form:

$$\underline{\underline{A}} = \begin{pmatrix} -\lambda & \lambda \\ \mu & -\mu \end{pmatrix}$$

Example 1.6

System with N identical components and N repairmen available.

Consider N identical components which can be in two states, working and failed. Let us assume that the components failures and repairs occur exponentially in time, with constant rates λ and μ, respectively. The states of the system can denote the number of failed components, i.e.:
State 0: none failed, all components function;
State 1: one component failed, N-1 function;
State 2: two components failed, N-2 function;
...
State N: all components failed, none functions.

The Markov graph is then:

Fig. 1.10

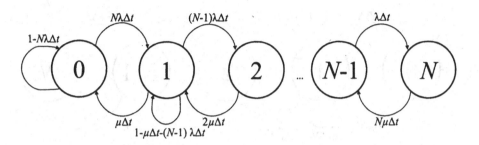

Given that only one event (failure or repair of one component) can occur in the small Δt and that the events are mutually exclusive, the value $N\lambda\Delta t$ is the probability that anyone of the N components fails in Δt. Similarly, $(N-k)\lambda\Delta t$ is the probability that anyone of the $(N-k)$ functioning components fails in Δt, whereas $k\mu\Delta t$ is the probability that anyone of the k failed components is repaired in Δt.

Example 1.7

System with N maintainable and identical components with only one repairman available.

Consider N identical components which can be in two states, working and failed. Let λ and μ be their constant failure and repair rates, respectively. The states of the system are the same as in the previous example. The Markov graph becomes:

Fig. 1.11

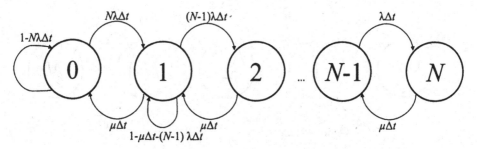

In this case with only one repairman available, the probability of repair in Δt is always $\mu\Delta t$ because only one component can be repaired.

1.3.2 Solution to the fundamental equation of the Markov process continuous in time

As seen in the previous Section, the system of linear, first-order differential equations in the state probabilities at time t, $P_j(t)$, $j = 0,1,2,...,N$, which governs the Markov process continuous in time is written in matrix form as:

$$\frac{d\underline{P}}{dt} = \underline{P}(t)\cdot\underline{\underline{A}} \qquad (1.58)$$

$$\underline{\underline{A}} = \begin{pmatrix} -\sum\limits_{j=1}^{N}\alpha_{0j} & \alpha_{01} & ... & \alpha_{0N} \\ \alpha_{10} & -\sum\limits_{\substack{j=0 \\ j\neq 1}}^{N}\alpha_{1j} & ... & \alpha_{1N} \\ ... & ... & ... & ... \end{pmatrix} \qquad (1.59)$$

This system is to be solved starting from the initial condition $\underline{P}(0) = \underline{C}$. The easiest method of solution is by Laplace transform. The Laplace transform of the state probability $P_j(t), j = 0,1,2,...,N$, denoted by $\tilde{P}_j(s)$, is defined as $\tilde{P}_j(s) = L[P_j(t)] = \int_0^\infty e^{-st} P_j(t) dt$; correspondingly, the Laplace transform of the time derivative of $P_j(t)$ is:

$$L\left(\frac{dP_j(t)}{dt}\right) = s \cdot \tilde{P}_j(s) - P_j(0), \quad j = 0,1,...,N \tag{1.60}$$

Laplace-transforming the fundamental equation of the Markov process (1.58):

$$s\underline{\tilde{P}}(s) - \underline{C} = \underline{\tilde{P}}(s) \cdot \underline{\underline{A}} \tag{1.61}$$

from which:

$$\underline{\tilde{P}}(s) = \underline{C} \cdot \left[s \cdot \underline{\underline{I}} - \underline{\underline{A}}\right]^{-1} \tag{1.62}$$

where $\underline{\underline{I}}$ is the identity matrix. Then, applying the inverse Laplace transformation we can retrieve the state probabilities vector $\underline{P}(t)$.

Furthermore, since the problems we deal with are all finite chains, there exists at least one recurrent state and thus, there is a steady state distribution $\underline{\Pi}$. This latter can be simply found by setting to zero the derivative of $\underline{P}$ in the fundamental equation:

$$\underline{\Pi} \cdot \underline{\underline{A}} = 0 \tag{1.63}$$

Taking into account that $\sum_{j=0}^{N} \Pi_j = 1$, the steady state probabilities are found to be equal to:

$$\Pi_j = \frac{D_j}{\sum_{i=0}^{N} D_i} \quad j = 0,1,2,...,N \tag{1.64}$$

where D_j is the determinant of the square matrix obtained from $\underline{\underline{A}}$ by deleting the j-th row and column.

Example 1.8

One component/one repairman with exponential failure and repair times distributions.

Consider one repairable component which can be in only two states, working (0) and failed (1), and one repairman. We assume that the component transition times are exponentially distributed with failure rate λ and repair rate μ. From Example 1.5, we know that the transition rate matrix has the form:

$$\underline{\underline{A}} = \begin{pmatrix} -\lambda & \lambda \\ \mu & -\mu \end{pmatrix}$$

The component is in operation at time $t=0$ ($\underline{C} = \begin{bmatrix} 1 & 0 \end{bmatrix}$). To compute the transient behaviour of the state probability vector we have to solve:

$$\underline{\tilde{P}}(s) = \underline{C} \cdot \left(s\underline{\underline{I}} - \underline{\underline{A}} \right)^{-1}$$

Thus, we need to compute the inverse matrix $\left(s\underline{\underline{I}} - \underline{\underline{A}} \right)^{-1}$:

$$\left(s\underline{\underline{I}} - \underline{\underline{A}} \right)^{-1} = \begin{pmatrix} s+\lambda & -\lambda \\ -\mu & s+\mu \end{pmatrix}^{-1}$$

$$= \frac{1}{\det\left[\left(s\underline{\underline{I}} - \underline{\underline{A}} \right)^{-1} \right]} \begin{pmatrix} s+\lambda & \lambda \\ \mu & s+\mu \end{pmatrix}$$

$$= \frac{1}{s^2 + \lambda s + \mu s} \cdot \begin{pmatrix} s+\lambda & \lambda \\ \mu & s+\mu \end{pmatrix}$$

From which we get:

$$\underline{\tilde{P}}(s) = \left(\frac{s+\lambda}{s \cdot (s+\lambda+\mu)} \quad \frac{\lambda}{s \cdot (s+\lambda+\mu)} \right)$$

Observing that the roots are 0 and $-(\lambda+\mu)$ and applying the inverse Laplace transformation, the state probability vector in the time domain becomes:

$$\underline{P}(t) = \left(\frac{\mu}{\lambda+\mu} + \frac{\lambda}{\lambda+\mu} \cdot e^{-(\lambda+\mu) \cdot t} \quad \frac{\lambda}{\lambda+\mu} - \frac{\lambda}{\lambda+\mu} \cdot e^{-(\lambda+\mu) \cdot t} \right)$$

where

$$P_0(t) = \frac{\mu}{\lambda+\mu} + \frac{\mu}{\lambda+\mu} \cdot e^{-(\lambda+\mu) \cdot t}$$

is the system instantaneous availability at time t (probability of being in operational state 0 at time t)

$$P_1(t) = \frac{\lambda}{\lambda+\mu} - \frac{\lambda}{\lambda+\mu} \cdot e^{-(\lambda+\mu) \cdot t}$$

is the system instantaneous unavailability at time t (probability of being in failed state 1 at time t).

The system steady state probabilities are readily found to be:

$$\Pi_0 = \frac{\mu}{\lambda+\mu} = \frac{\frac{1}{\lambda}}{\frac{1}{\mu}+\frac{1}{\lambda}} = \frac{MTBF}{MTTR+MTBF} = \text{average fraction of time the}$$

system is functioning

$$\Pi_1 = \frac{\lambda}{\lambda + \mu} = \frac{\dfrac{1}{\mu}}{\dfrac{1}{\mu} + \dfrac{1}{\lambda}} = \frac{MTTR}{MTTR + MTBF} = \text{average fraction of time the}$$

system is down (under repair)

where MTBF and MTTR are the Mean Time Between Failures and the Mean Time To Repair, respectively.

1.3.3 Failure Intensity

The unconditional probability of arriving in state j in the next Δt departing from state i at time t is given by:

$$
\begin{aligned}
P\big[X(t+\Delta t) &= j, X(t) = i\big] \\
&= P\big[X(t+\Delta t) = j \,\big|\, X(t) = i\big] \cdot P\big[X(t) = i\big] \qquad (1.65) \\
&= p_{ij}(\Delta t) \cdot P_i(t)
\end{aligned}
$$

Then, the frequency of departure from state i to state j is:

$$v_{ij}^{dep}(t) = \lim_{\Delta t \to 0} \frac{p_{ij}(\Delta t) \cdot P_i(t)}{\Delta t} = \alpha_{ij} \cdot P_i(t) \qquad (1.66)$$

which at steady state becomes:

$$v_{ij}^{dep} = \alpha_{ij} \cdot \Pi_i \qquad (1.67)$$

The total frequency of departure from state i to any other state j is then:

$$v_i(t) = \sum_{\substack{j=0 \\ j \neq i}}^{N} \alpha_{ij} \cdot P_i(t) = \alpha_{ii} \cdot P_i(t) \qquad (1.68)$$

and at steady state:

$$v_i = \alpha_{ii} \cdot \Pi_i \qquad (1.69)$$

Similarly, we can consider arrivals to state i from any state k and define:

$$v_i^{arr}(t) = \sum_{\substack{k=0 \\ k \neq i}}^{N} \alpha_{ki} \cdot P_k(t)$$

$$v_i^{arr} = \sum_{\substack{k=0 \\ k \neq i}}^{N} \alpha_{ki} \cdot \Pi_k \qquad (1.70)$$

Since the matrix equation $\underline{\Pi} \cdot \underline{\underline{A}} = 0$ implies

$$\alpha_{ii} \cdot \Pi_i = \sum_{\substack{k=0 \\ k \neq i}}^{N} \alpha_{ki} \cdot \Pi_k \qquad i = 0,1,2,...,N \qquad (1.71)$$

at steady state the frequency of departure from state i, v_i, is equal to the frequency of arrivals to state i, v_i^{arr}.

In our case of interest we define the system failure intensity W_f as the rate at which system failures occur, i.e. the expected number of system failures per unit of time; this is equivalent to the rate of exiting a success state to go into one of fault. Denoting by S an F the sets of success and fault states of the system, respectively:

$$W_f(t) = \sum_{i \in S} P_i(t) \cdot \lambda_{i \to F} \qquad (1.72)$$

where the sum is over all the success states $i \in S$ of the system, $P_i(t)$ is the probability of the system being in the functioning state i at time t and $\lambda_{i \to F}$ is the conditional probability of leaving the state i of success towards a failed state.

Similarly, we define the system repair intensity W_r as the rate at which system repairs occur, i.e. the rate of exiting a failed state to return into a success state:

$$W_r(t) = \sum_{j \in F} P_j(t) \cdot \mu_{j \to s} \qquad (1.73)$$

where the sum is over all the failed states $j \in F$ of the system, $P_j(t)$ is the probability of the system being in the failed state j at time t and $\mu_{j \to s}$ is the conditional probability of leaving the state j of failure towards a functioning state.

Example 1.9

One component/one repairman with exponential failure and repair times distributions.

In this case, from the results of Example 1.8, the system failure and repair intensities are computed simply as follows:

$$W_f(t) = \lambda \cdot P_0(t)$$
$$W_r(t) = \mu \cdot P_1(t)$$

1.3.4   Average time of occupancy of a given state *i*

When the process arrives at state *i*, the system will remain in such state a time T_i before it departs towards another state with rate:

$$\alpha_{ii} = \sum_{\substack{j=0 \\ j \neq i}}^{N} \alpha_{ij} \qquad i = 0,1,2,3,...,N$$

Since the departure rate α_{ii} is constant, the duration T_i of occupancy of state *i* is exponentially distributed with parameter α_{ii} and the mean duration of stay in state *i* is:

$$l_i = \frac{1}{\alpha_{ii}}$$

Thus, from the steady state relation $v_i = \alpha_{ii} \cdot \Pi_i$ (Eq. 1.70) we have:

$$v_i = \alpha_{ii} \cdot \Pi_i = \frac{\Pi_i}{l_i}$$

and

$$\Pi_i = v_i \cdot l_i$$

The mean proportion of time Π_i that the system spends in state i is equal to the visit frequency to state i multiplied by the mean duration of one visit in state i.

1.3.5 System availability

Among all possible states of the system, some will represent the system functioning properly, according to some specified criteria of system performance, whereas others will denote configurations in which the system is failed. As before, let S denote the subset of states in which the system is functioning and F the subset of failed states.

The system instantaneous availability at time t is computed simply by summing the probabilities of being in a success state at time t:

$$p(t) = \sum_{i \in S} P_i(t) = 1 - q(t) = 1 - \sum_{j \in F} P_j(t) \qquad (1.74)$$

In the Laplace domain we write:

$$\tilde{p}(s) = \sum_{i \in S} \tilde{P}_i(s) = \frac{1}{s} - \sum_{j \in F} \tilde{P}_j(s) \qquad (1.75)$$

1.3.6 System reliability

We distinguish two cases depending on whether repairs are allowed or not.

System unattended (no repairs allowed)

In the case of unattended systems, repairs cannot be performed and the system reliability coincides with its availability [7]:

$$R(t) \equiv p(t) = 1 - q(t) \qquad (1.76)$$

Hence, in the Laplace domain, from $\underline{\tilde{P}}(s)$ we can simply find those elements $\tilde{P}_i(s)$ which correspond to system success states $i \in S$ and write analogously to (1.75):

$$\tilde{R}(s) = \sum_{i \in S} \tilde{P}_i(s) = \frac{1}{s} - \sum_{j \in F} \tilde{P}_j(s) \qquad (1.77)$$

The mean-time-to-failure, MTTF, can then be computed as (exploiting the proprieties of the Laplace transform):

$$MTTF = \int_0^{\infty} R(t)\,dt = \sum_{i \in S} \tilde{P}_i(0) = \tilde{R}(0) = \left[\frac{1}{s} - \sum_{j \in F} \tilde{P}_j(s) \right]_{s=0} \qquad (1.78)$$

Attended system (repairs allowed)

In this case, the following procedure must be performed to compute the system reliability:

1. Partition the transition rate matrix $\underline{\underline{A}}$ so as to exclude all failed states $j \in F$, which are now considered absorbing. The resulting matrix $\underline{\underline{A}}'$ is the matrix of the transition rates for transitions only

among the "success states" $i \in S$. As long as the system switches back and forth within the states of $\underline{\underline{A}}'$, it is functioning continuously with no interruption.

2. Solve the reduced problem of $\underline{\underline{A}}'$ for the probabilities $P_i^*(t)$, $i \in S$, of being in these (transient) safe states. The reliability is then easily computed by summing all such probabilities:

$$R(t) = \sum_{i \in S} P_i^*(t) \qquad (1.79)$$

and from (1.78) the mean-time-to-failure, MTTF, is:

$$MTTF = \int_0^\infty R(t)\, dt = \sum_{i \in S} \tilde{P}_i^*(0) = \tilde{R}(0) \qquad (1.80)$$

Note that at steady state all the state probabilities, $P_i^*(\infty)$, are equal to zero because in this modified (reduced) problem, we have only transient states.

Example 1.10

System with two identical components which can fail with constant rate λ and be repaired by two repairmen at constant rate μ.

The Markov diagram of the process is:

Fig. 1.12

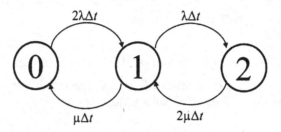

and the transition matrix:

$$\underline{\underline{A}} = \begin{pmatrix} -2\lambda & 2\lambda & 0 \\ \mu & \mu+\lambda & \lambda \\ 0 & 2\mu & -2\mu \end{pmatrix}$$

To answer any system availability or reliability question, we must specify the system logic of operation.

a) Parallel logic (1 out of 2). The states of the system are:

State 0: system is operating (both components functioning)

State 1: system is operating (only one of the two components functioning)

State 2: system is failed (both components failed)

The system reliability at a given time is the probability of the system being in states 0 or 1 continuously from $t=0$, i.e. accounting for the fact that it cannot come back from state 2 which is an absorbing state with respect to the reliability measure. Thus, the partition of the transition rate matrix is done including only the success states 0 and 1:

$$\underline{\underline{A}} = \left(\begin{array}{cc|c} -2\lambda & 2\lambda & 0 \\ \mu & -(\mu+\lambda) & \lambda \\ \hline 0 & 2\mu & -2\mu \end{array} \right) \Rightarrow \underline{\underline{A}}' = \begin{pmatrix} -2\lambda & 2\lambda \\ \mu & -(\mu+\lambda) \end{pmatrix} \qquad (1.81)$$

The fundamental equation for the state probabilities of the reduced problem is:

$$\frac{d\underline{P}^*}{dt} = \underline{P}^*(t) \cdot \begin{pmatrix} -2\lambda & 2\lambda \\ \mu & -(\lambda+\mu) \end{pmatrix} \qquad (1.82)$$

with initial condition $\underline{P}^*(0) = (1 \ 0)$.
Laplace-transforming:

$$\underline{\tilde{P}}^*(s) = (1 \ 0) \cdot \left(s\underline{\underline{I}} - \underline{\underline{A}}'\right)^{-1} \qquad (1.83)$$

with

$$\left(s\underline{\underline{I}} - \underline{\underline{A}}'\right)^{-1} = \frac{1}{(s-\omega_0)(s-\omega_1)} \begin{pmatrix} s+\lambda+\mu & 2\lambda \\ \mu & s+2\lambda \end{pmatrix} \qquad (1.84)$$

The determinant roots are:

$$\omega_{0,1} = \frac{-3\lambda - \mu \pm \sqrt{\lambda^2 + 6\lambda\mu + \mu^2}}{2} \qquad (1.85)$$

and the inverse-transformed reliability is:

$$R(t) = \frac{\omega_0 \cdot e^{\omega_1 t} - \omega_1 \cdot e^{\omega_0 t}}{\omega_0 - \omega_1} \qquad (1.86)$$

The mean time to failure (MTTF) can be computed as follows:

$$MTTF = \tilde{R}(0) = \sum_i \tilde{P}_i^*(0) \qquad (1.87)$$

Since $\underline{\tilde{P}}^*(s) = \underline{C}^* \cdot \left(s \cdot \underline{\underline{I}} - \underline{\underline{A}}'\right)^{-1}$ and introducing the unit vector $\underline{w} = \begin{bmatrix} 1 & 1 & 1 & \dots & 1 \end{bmatrix}^T$,

$$MTTF = \underline{C}^* \cdot \left(-\underline{\underline{A'}}\right)^{-1} \cdot \underline{w}^T \qquad (1.88)$$

In our specific case, we have:

$$
\begin{aligned}
MTTF &= \begin{pmatrix} 1 & 0 \end{pmatrix} \cdot \begin{pmatrix} 2\lambda & -2\lambda \\ -\mu & \mu+\lambda \end{pmatrix}^{-1} \cdot \begin{pmatrix} 1 \\ 1 \end{pmatrix} \\
&= \begin{pmatrix} 1 & 0 \end{pmatrix} \cdot \frac{1}{2\lambda(\lambda+\mu)-2\lambda\mu} \begin{pmatrix} \mu+\lambda & 2\lambda \\ \mu & 2\lambda \end{pmatrix} \cdot \begin{pmatrix} 1 \\ 1 \end{pmatrix} = \\
&= \frac{1}{2\lambda^2}\begin{pmatrix} \mu+\lambda & 2\lambda \end{pmatrix} \cdot \begin{pmatrix} 1 \\ 1 \end{pmatrix} = \frac{3\lambda^2+\mu}{2\lambda^2} = \\
&= \frac{3}{2\lambda} + \frac{\mu}{2\lambda^2}
\end{aligned}
\qquad (1.89)
$$

We can also compute the failure intensity (Eq. 1.73):

$$W_f = P_1(t) \cdot \lambda \qquad (1.90)$$

and the repair intensity (Eq. 1.74):

$$W_r(t) = P_2(t) \cdot 2\mu \qquad (1.91)$$

b)   Series logic (2 out of 2). The states of the system are:

State 0: system is operating (both components functioning)

State 1: system is failed (only one component functioning and the other failed)

State 2: system is failed (both components failed)

The system reliability at a given time is the probability of the system being in state 0 continuously from $t=0$, i.e. accounting for the fact that it cannot come back from states 1 or 2, which are absorbing. Thus, the partition of the transition rate matrix, including only success states, is:

$$
\underline{\underline{A}} = \begin{pmatrix} -2\lambda & 2\lambda & 0 \\ \mu & -(\lambda+\mu) & \lambda \\ 0 & 2\mu & -2\mu \end{pmatrix} \Rightarrow \underline{\underline{A}}' = -2\lambda \tag{1.92}
$$

In this case, it is easy to solve the reduced problem directly in the time domain:

$$
\frac{d\underline{P}^*}{dt} = \underline{P}^* \cdot \underline{\underline{A}}'; \ \underline{P}^*(0) = \underline{C}^* \tag{1.93}
$$

which simplifies to:

$$
\frac{dP_0^*}{dt} = -2\lambda \cdot P_0^*
$$
$$
P_0^*(t) + P_1^*(t) = 1 \tag{1.94}
$$
$$
P_0^*(0) = 1
$$
$$
P_1^*(0) = 0
$$

that leads to the solution:

$$
P_0^*(t) = e^{-2\lambda t} \tag{1.95}
$$

which is the probability of both independent exponential components being functioning with no failures up to time t.

Other quantities of interest with respect to the system reliability characteristics are:

Steady-state failure intensity

The steady-state intensity W_f is defined as the expected number of visits to (arrivals into) a failed state $i \in F$ per unit time, computed over a long period of time.

Mean duration of system failure

The mean duration l_f of a system failure is defined as the mean time from when the system enters into a failed state ($i \in F$) until it is repaired/restored and brought back into a functioning state (S). From $\Pi_i = v_i \cdot l_i$ it is obvious that the system steady-state unavailability $q_\infty = 1 - p_\infty$ [7] is equal to the frequency of system failures multiplied by the mean duration of system failure. Hence

$$q_\infty = W_f \cdot l_f \qquad (1.96)$$

Mean time between system failures

The mean time between system failures MTBF is the mean time between consecutive transitions from a functioning state ($i \in S$) into a failed state ($j \in F$). The MTBF may be computed from the steady state frequency of system failures by:

$$MTBF = \frac{1}{W_f} \qquad (1.97)$$

Example 1.11 [8]

Spare parts modelling

This example is intended to show how a Markov model can be built to catch the stochastic dynamics of a plant supported by spare parts.

The Markov system representation is based upon three indexes: the first one indicates the number of operating components in the system; the second, the number of spares available in storage; the third one is the number of units in recycling (i.e. in the repair facility).

Consider for simplicity a one-unit system supported by one single spare and one repair facility: a Markov model can be built for the complete recycling process of these two-components system: upon failure, the component is replaced by the spare unit and sent for repair at a remote facility. Upon repair, it is shipped back and serves as a spare. The recycling cycle is portrayed in Fig. 1.13 and the possible system states are reported in Table 1.1. The governing stochastic processes are assumed exponential in time.

Fig. 1.13 Operating unit supported by spare and repair facility

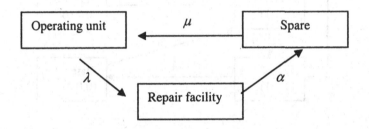

Table 1.1 System states

System state	Operating units	Spare units	Units under repair
1	1	1	0
2	1	0	1
3	0	1	1
4	0	2	0
5	0	0	2

The initial system state (state 1) is the nominal one (1,1,0), i.e. the online component is operational and the single spare is available. From this state, upon failure of the operating component the system transfers into state 3 (0,1,1), in which the failed unit is sent to repair. The rate at which this transfer occurs is λ. From the state (0,1,1), the system can move into state 2 (1,0,1), at a rate μ, if the spare unit is sent into operation before the failed unit is returned from repair, or into state 4 (0,2,0), at a rate α, if the failed unit returns from repair before the spare is started up. From state 2, transfer is possible into state 5 (0,0,2) at a rate λ, or into the nominal state (1,1,0) at a rate α. From state 4 transfer is possible only into the nominal state, at a rate μ. From state 5, transfer is possible only to state 3, at a rate 2α. Fig. 1.14 reports the system state transitions.

Fig. 1.14 Markov diagram for a one-unit system supported by a single spare and repair facility

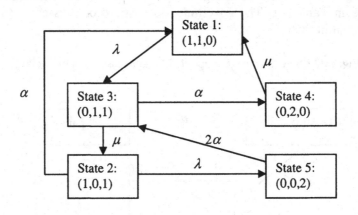

The set of equations governing the Markov model is:

$$\frac{dP_1(t)}{dt} = -\lambda P_1(t) + \alpha P_2(t) + \mu P_4(t)$$

$$\frac{dP_2(t)}{dt} = -(\alpha + \lambda)P_2(t) + \mu P_3(t)$$

$$\frac{dP_3(t)}{dt} = -(\alpha + \mu)P_3(t) + \lambda P_1(t) + 2\alpha P_5(t)$$

$$\frac{dP_4(t)}{dt} = -\mu P_4(t) + \alpha P_3(t)$$

$$\frac{dP_5(t)}{dt} = -2\alpha P_5(t) + \lambda P_2(t)$$

with the initial condition $P_l(0) = \delta_{l,1}$, $l = 1, 2, ..., 5$, where $\delta_{l,1}$ is the usual Kronecker delta function equal to 1 for $l=1$ (i.e., the system is in state 1 at the initial time).

From the knowledge of the state probabilities $P_l(t), l = 1, 2, ..., 5$ it is easy to compute the reliability and availability measures of interest, as explained in Sections 1.3.5 and 1.3.6.

When we move to consider a system of N_c identical, exponential components supported by N_{sp} spares, the number of possible system states increases remarkably, so that a complete description of the recycling process becomes analytically impracticable.

As a possible simplification let us go back to consider that the replacement time is negligible with respect to the repair and failure mean times. In this case, with reference to the system states of Table 1.2 and the Markov diagram of Fig. 1.15, the initial, nominal state is denoted by $(N_c, N_{sp}, 0)$ and can transfer, at a rate $N_c\lambda$, only into state 2, denoted by $(N_c, N_{sp}-1, 1)$, in which the failed unit is instantaneously replaced by a spare and sent to repair. From state 2 , the system can return into state 1 at a rate α if the repair of the failed units occurs, or transfer into state number 3 $(N_c, N_{sp}-2, 2)$, at a rate $N_c\lambda$, upon an additional unit failure. Similarly, the first N_c+1 states are of the form $(N_c, N_{sp}-j, j)$, with $j=0,1,...,N_{sp}$. The rate of flow from state j into state $j+1$ is $N_c\lambda$, whereas the return rate is $j\alpha$. All these states are characterized by the fact that they all have N_c operating components, the last state in this group being $(N_c, 0, N_{sp})$. Then, there are N_c additional states for which no spares are available which take the form $(N_c-i, 0, N_{sp}+i)$, with $i=1,2,...,N_c$. Now, transfer is possible at a rate $(N_{sp}+i)\alpha$, to state $(N_c-i+1, 0, N_{sp}+i-1)$ upon repair of a failed unit (followed by its instantaneous start up) or to state $(N_c-i-1, 0, N_{sp}+i+1)$, at a rate $\lambda(N_c-i)$, because of the failure of an on-line component.

Table 1.2: System states

System state	Operating units	Spare units	Units under repair
1	N_c	N_{sp}	0
2	N_c	$N_{sp}-1$	1
3	N_c	$N_{sp}-2$	2
...	...	...	...
$j+1$	N_c	$N_{sp}-j$	j
...	...	...	...
$N_{sp}+1$	N_c	0	N_{sp}
$N_{sp}+2$	N_c-1	0	$N_{sp}+1$
$N_{sp}+3$	N_c-2	0	$N_{sp}+2$
...	...	...	...
$N_{sp}+i+1$	N_c-i	0	$N_{sp}+i$
...	...	...	...
$N_{sp}+N_c+1$	0	0	$N_{sp}+N_c$

Fig. 1.15 Markov diagram for a system with N_c identical repairable components supported by N_{sp} spares and negligible replacement time

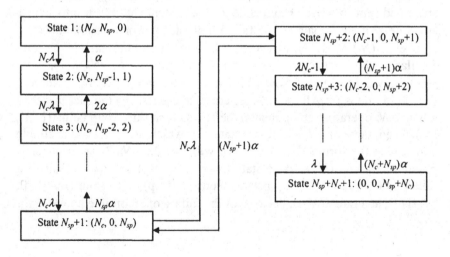

Example 1.12 [9]

Modelling 'on condition' maintenance strategies of a deteriorating component.

In this Example we describe a Markov approach for modelling the behaviour of a deteriorating component subject to condition-based preventive maintenance.

Let $t=\{t_0,t_1,...,T_M\}$ represent the discretized time variable and $X=\{x_0,x_1,...,x_m,x_{m+1}\}$ be a discrete random variable denoting the level of degradation of the component. The process of degradation evolution is described through the first $m+1$ states $(x_0,x_1,...,x_m)$, while the state x_{m+1} refers to a non reparable degradation condition, reachable upon a possible random failure occurring to the component while in any of the other operative states $x_i < x_{m+1}$. We define:

$P_n(k)$ = probability of being in degradation level x_k at time t_n;

$q(j|k)$ = probability of an increase of j units of degradation starting from an initial degradation level of k units (thus, the final degradation state will be of $k+j$ units);

$r(j|k)$ = probability of a decrease of j units of degradation starting from an initial degradation level of k units, due to possible repair of the component (thus, the final degradation level will be of $k-j$ units);

k_{th} = threshold degradation level beyond which maintenance or repair action is required;

$f(k)$ = probability of component failure due to a random shock which leads the component from the current degradation level, k, to the final degradation state, x_{m+1}.

The above defined probabilities allow us to describe the physics of the evolution of a component through its states of operation. A possible realization of the time evolution of a component is shown in Fig. 1.16.

Fig. 1.16 Time sketch of the degradation process of a generic component

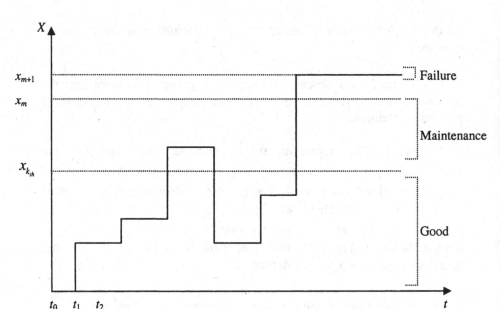

On the basis of the above definitions, a Markov model for the process of degradation and 'on-condition' maintenance can be built. Initially, we do not consider the effects of possible random failures occurring to the component, i.e. the component state evolves only through the first $m+1$ (x_0, x_m) degradation or maintenance states without failing due to random shocks. Obviously, when the component is operating (degradation state $k \leq k_{th}$) only degradation increases are possible, whereas when the component is under maintenance (degradation state $k > k_{th}$) only degradation recoveries can occur. The process is governed by the following system of equations (in order to simplify the notations $q(j|k)$ and $r(j|k)$ are replaced with q_{jk} and r_{jk}):

$$
\begin{pmatrix} P_n(0) \\ P_n(1) \\ P_n(2) \\ P_n(3) \\ \dots \\ P_n(k_{th}) \\ P_n(k_{th}+1) \\ \dots \\ P_n(m) \end{pmatrix}
=
\begin{pmatrix}
q_{00} & 0 & 0 & \dots & 0 & r_{k_{th}+1\,k_{th}+1} & r_{k_{th}+2\,k_{th}+2} & \dots & r_{mm} \\
q_{10} & q_{01} & 0 & \dots & 0 & r_{k_{th}\,k_{th}+1} & r_{k_{th}+1\,k_{th}+2} & \dots & r_{m-1\,m} \\
q_{20} & q_{11} & q_{02} & \dots & 0 & r_{k_{th}-1\,k_{th}+1} & r_{k_{th}\,k_{th}+2} & \dots & r_{m-2\,m} \\
q_{30} & q_{21} & q_{12} & \dots & 0 & r_{k_{th}-2\,k_{th}+1} & \dots & \dots & r_{m-3\,m} \\
\dots & \dots & \dots & \dots & \dots & \dots & \dots & \dots & \dots \\
q_{k_{th}\,0} & q_{k_{th}-1\,1} & q_{k_{th}-2\,2} & \dots & q_{0\,k_{th}} & r_{1\,k_{th}+1} & r_{2\,k_{th}+2} & \dots & r_{m-k_{th}\,m} \\
q_{k_{th}+1\,0} & q_{k_{th}\,1} & q_{k_{th}-1\,2} & \dots & q_{1\,k_{th}} & \dots & 0 & \dots & 0 \\
\dots & \dots & \dots & \dots & \dots & \dots & 0 & \dots & 0 \\
q_{m\,0} & q_{m-1\,1} & q_{m-2\,2} & \dots & q_{m-k_{th}\,k_{th}} & \dots & 0 & \dots & 0
\end{pmatrix}
\cdot
\begin{pmatrix} P_{n-1}(0) \\ P_{n-1}(1) \\ P_{n-1}(2) \\ P_{n-1}(3) \\ \dots \\ P_{n-1}(k_{th}) \\ P_{n-1}(k_{th}+1) \\ \dots \\ P_{n-1}(m) \end{pmatrix}
\tag{1.98}
$$

As for the probabilities $q(j|k)$, they are defined only for values of the starting degradation level $k \le k_{th}$ and small degradation increments are favoured among the feasible values $j = 0, 1, ..., m - k$. Thus, we arbitrarily set:

$$q(j|k) = \alpha_k \left(1 - \frac{j}{N-k}\right);$$

$N = m + 1 =$ total number of levels of degradation
$k = 0, 1, ..., k_{th}; \quad j = 0, 1, ..., N - 1 - k$

The value of the coefficient α_k in the previous equation is obtained from the normalization of the probabilities:

$$\sum_{j=0}^{m-k} q(j|k) = 1 \qquad k = 0, 1, 2, .., k_{th};$$

And the expression for the probabilities $q(j|k)$ becomes:

$$q(j|k) = \frac{2}{N-k+1} \cdot \frac{N-k-j}{N-k} \qquad \text{for } k = 0, 1, ..., k_{th}; \ j = 0, 1, ..., N - 1 - k$$

Fig. 1.17 shows the values of the probabilities $q(j|k)$ as a function of the degradation increment j, for three different values of k. The total number of degradation levels is $N = 71$. The probabilities $q(j|k)$

decrease linearly with j, so that small degradation increments are favoured. Also, the slope of $q(j|k)$ increases in absolute value with the initial degradation level k, so that small increments are more and more favored as the degradation proceeds.

Fig. 1.17 Values of the probabilities $q(j|k)$ as a function of the degradation increment j, for three different values of the starting degradation level k (total number of degradation levels $N = 71$)

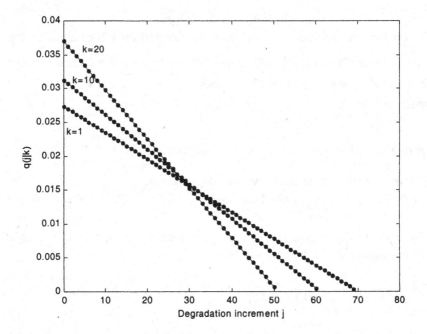

With regards to the values of the repair probabilities $r(j|k)$, we choose to favor the transitions with higher values of degradation recovery, i.e. repairs which effectively reduce degradation. Indeed, we assume that a repair can induce a decrease in the degradation which ranges from a minimum value k_{min}, here assumed to be equal to $k-k_{th}/2$ (corresponding to a final degradation level of $k_{th}/2$ units) to a maximum value k, when complete repair to an 'as good as new' state is performed:

$$r(j|k) = \beta_k \cdot j \quad \text{for } k_{min} \leq j \leq k; \quad k > k_{th}; \quad k_{min} = k - k_{th}/2$$
$$r(j|k) = 0 \quad \text{for } j < k_{min}$$

where the coefficients β_k are found from the normalization condition:

$$\sum_{j=k_{min}}^{k} r(j|k) = 1$$

leading to:

$$\beta_k = \frac{2}{k(k+1) - k_{min}(k_{min} - 1)} \quad k > k_{th}$$

Fig. 1.18 reports the values of the probabilities $r(j|k)$ as a function of the degradation recovery j starting from a degradation level of k units, for three different values of k. Note that the probabilities $r(j|k)$ are taken such that the lower the initial degradation level is, the steeper the slope of $r(j|k)$ is, thus favoring degradation recoveries closer to the 'as good as new' condition.

Fig. 1.18 Values of the probabilities $r(j|k)$ **as a function of the degradation recovery j, for three different values of the starting degradation level k ($N =$ 71, $k_{th}= 30$, $k_{min}=k$ -15)**

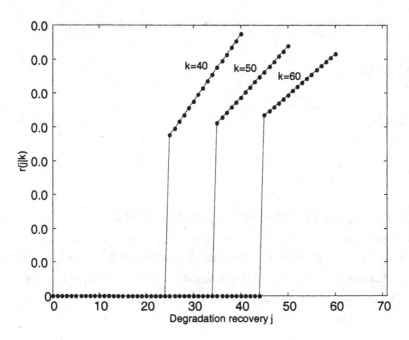

The system of equations (1.98) is intended to describe the behaviour of a component which evolves through degradation and repair. We now associate to each level of degradation a probability of shock failure which will realistically increase as the component degradation increases. Thus, we further define:

$f(k)=$ probability of component failure when at degradation level k

For modeling the effect of the failures, we proceed as follows. We consider absorbing the state x_{m+1}, which can be reached upon component failure from states with degradation level $k \leq k_{th}$, as it is assumed that the component cannot fail while under repair. From each operating state k, the component can either fail, i.e. transfer from state x_k to state x_{m+1}, with probability $f(k)$, or increase its degradation level of j units, with probability $q(j|k)\cdot(1-f(k))$, since the two events of failure and degradation are mutually exclusive. The component's evolution is now governed by the following system of equations:

$$
\begin{pmatrix} P_n(0) \\ P_n(1) \\ P_n(2) \\ P_n(3) \\ \cdots \\ P_n(k_{th}) \\ P_n(k_n+1) \\ \cdots \\ P_n(m) \\ P_n(m+1) \end{pmatrix} =
\begin{pmatrix}
q_{00}(1-f(0)) & 0 & 0 & \cdots & 0 & r_{k_{th}+1\,k_{th}+1} & r_{k_{th}+2\,k_{th}+2} & \cdots & r_{mm} & 0 \\
q_{10}(1-f(0)) & q_{01}(1-f(1)) & 0 & \cdots & 0 & r_{k_{th}\,k_{th}+1} & r_{k_{th}+1\,k_{th}+2} & \cdots & r_{m-1\,m} & 0 \\
q_{20}(1-f(0)) & q_{01}(1-f(1)) & q_{02}(1-f(2)) & \cdots & 0 & r_{k_{th}-1\,k_{th}+1} & r_{k_{th}\,k_{th}+2} & \cdots & r_{m-2\,m} & 0 \\
q_{30}(1-f(0)) & q_{01}(1-f(1)) & q_{12}(1-f(2)) & \cdots & 0 & r_{k_{th}-2\,k_{th}+1} & r_{k_{th}-1\,k_{th}+2} & \cdots & r_{m-3\,m} & 0 \\
\cdots & \cdots & \cdots & & \cdots & \cdots & \cdots & \cdots & \cdots & 0 \\
q_{k_{th}0}(1-f(0)) & q_{k_{th}-1\,1}(1-f(1)) & q_{k_{th}-2\,2}(1-f(2)) & \cdots & q_{0k_{th}}(1-f(k_{th})) & r_{1\,k_{th}+1} & r_{2\,k_{th}+2} & \cdots & r_{m-k_{th}\,m} & 0 \\
q_{k_{th}+1\,0}(1-f(0)) & q_{k_{th}\,1}(1-f(1)) & q_{02}(1-f(2)) & \cdots & q_{1k_{th}}(1-f(k_{th})) & 0 & 0 & \cdots & 0 & 0 \\
\cdots & & \cdots & & \cdots & & 0 & 0 & \cdots & 0 & 0 \\
q_{m0}(1-f(0)) & q_{m-1\,1}(1-f(1)) & q_{m-2\,2}(1-f(2)) & \cdots & q_{m-k_{th}\,k_{th}}(1-f(k_{th})) & 0 & 0 & \cdots & 0 & 0 \\
f(0) & f(1) & f(2) & \cdots & f(k_{th}) & 0 & 0 & \cdots & 0 & 1
\end{pmatrix}
\begin{pmatrix} P_{n+1}(0) \\ P_{n+1}(1) \\ P_{n+1}(2) \\ P_{n+1}(3) \\ \cdots \\ P_{n+1}(k_{th}) \\ P_{n+1}(k_{th}+1) \\ \cdots \\ P_{n+1}(m) \\ P_{n+1}(m+1) \end{pmatrix}
\quad (1.99)
$$

The component is unavailable if its degradation level k is above k_{th}, either because it is under repair or because it has failed ($k = m+1$). Thus, the instantaneous availability at time t_n of a component C, $A^C(t_n)$, can be obtained as:

$$
A^C(t_n) = \sum_{i=0}^{k_{th}} P_n(i)
$$

The probability $M^C(t)$ that at time t_n a maintenance action is being performed on component C is:

$$
M^C(t_n) = \sum_{i=k_{th}+1}^{m} P_n(i)
$$

When generalizing to systems with components of N_t different kinds, we need to extend the notations introduced:

$N_i = m_i + 1 =$ total number of levels of degradation, for the i-th component, $i=1,2,\ldots,N_C$;

$X^i = \{ x_0^i, x_1^i, \ldots, x_{m_i}^i \} =$ discrete random variable denoting the level of degradation of component i;

$P_n^i(k) =$ probability, for the i-th component, of being in degradation level x_k at time t_n;

$q^i(j|k)$ = probability, for the i-th component, of an increase of j units of degradation starting from an initial degradation level of k units (thus, the final degradation state will be of $k+j$ units);

$r^i(j|k)$ = probability of a decrease of j units of degradation starting from an initial degradation level of k units, due to maintenance or repair of the i-th component (thus, the final degradation level will be of $k-j$ units);

k_{th}^i = threshold degradation level beyond which a random decrease occurs following a maintenance or repair action, for component i;

$f^i(k)$ = probability of failure of component i, when the degradation level is k.

$$q_i(j|k) = \frac{2}{N_i - k + 1} \cdot \frac{N_i - k - j}{N_i - k}$$

$$\text{for } k = 0,1,...,k_{th}^i; \quad j = 0,1,...,N_i - 1 - k$$

$$r_i(j|k) = \frac{2}{k(k+1) - k_{min}^i (k_{min}^i - 1)} \cdot j$$

$$\text{for } k_{min}^i \leq j \leq k; \quad k > k_{th}^i; \quad k_{min}^i = k - k_{th}^i / 2$$

$$r_i(j|k) = 0 \text{ for } j < k_{min}^i.$$

Finally, for the i-th component, the availability at time t_n, $A^i(t_n)$, and the probability of being under maintenance, $M^i(t_n)$, can be obtained from the system of equations (1.99):

$$A^i(t_n) = \sum_{l=0}^{k_{th}} P_n^i(l)$$

$$M^i(t_n) = \sum_{l=k_{th}+1}^{m_i} P_n^i(l)$$

From the knowledge of the components' availabilities, the expression for the system availability $A(t)$ can then be simply determined considering the logic of the series-parallel system and the corresponding laws of probability. Once the instantaneous system availability is determined, we can compute the objective function, mean system availability $\overline{A}$ over the mission time T_M, from its definition [7].

1.4 References

[1] Rausand, M. and Hoyland, A., *System Reliability Theory*, John Wiley & Sons, 2004.

[2] Birolini, A., *Reliability Engineering: Theory and Practice*, Springer-Verlag, ISBN 3-540-40287-X, 2004.

[3] Ushakov, I. A., *Handbook of Reliability Engineering*, John Wiley & Sons, 1994.

[4] Limnios, N., Oprisan, G., *Semi-Markov Processes and Reliability*, Statistics for Industry and Technology, ISBN: 978-0-8176-4196-2, A Birkhäuser book, 2001.

[5] Barbu, V., Limnios, N., *Semi-Markov Chains and Hidden Semi-Markov Models toward Applications,* Lecture Notes in Statistics , Vol. 191, Springer, ISBN: 978-0-387-73171-1, 2008.

[6] Howard, R. A., *Dynamic Probabilistic Systems*, John Wiley & Sons, New York, 1971.

[7] Zio, E., An Introduction to the Basics of Reliability and Risk Analysis, *Series in Quality, Reliability and Engineering Statistics*, Vol.13, World Scientific, Singapore, 2007.

[8] M. Marseguerra, L. Podofillini, E. Zio, *Use of Genetic Algorithms for the Optimization of Spare Parts Inventory,* ESREL 2001, European Safety and Reliability, Torino (Italy), Sept 16-20, 2001, pp. 1523-1530.

[9] Marseguerra, M., Zio, E., Podofillini, L., *Condition Based Maintenance Optimization by Means of Genetic Algorithms and Monte Carlo Simulation*, Reliability Engineering and System Safety, Vol. 77 No.2, pp.151-65, 2002.

2. MONTE CARLO SIMULATION FOR RELIABILITY AND AVAILABILITY ANALYSIS

2.1 Introduction

The Monte Carlo (MC) simulation may be the only method that can yield solutions to complex multi-dimensional stochastic modeling problems such as those typically involved in reliability and availability analysis. For about three decades it was used almost exclusively, and extensively, in nuclear technology [1-4]. Presumably, the main reason for its use being limited to only nuclear applications was the lack of suitable computing power: indeed, the method is computer memory- and time-intensive. With the increasing availability of fast computers, the application of the method has become more and more feasible in the practice of various fields of system engineering [5-7].

This chapter introduces the basic principles underlying the MC method for the evaluation of the reliability and availability of complex systems. Most of the enclosed material is adapted from the book *"Basics of the Monte Carlo Method with Application to System Reliability"* by M. Marseguerra and E. Zio, LiLoLe-Verlag GmbH (Publ. Co. Ltd.), 2002, with the permission of the publisher.

2.2   Monte Carlo simulation for system engineering

Let us consider an engineering system made up of N_c physical components (pumps, valves, ducts, electronic circuitry, and so on). Each component can be in a number of states, e.g. working, failed, standby, etc. During its life, a component may move from one state to another by a transition which occurs stochastically in time and whose outcome (final state reached) is stochastic. The stochastic behaviour of each component is then defined by a matrix of probabilities of transition between different states.

On the other hand, the full description of the system stochastic behaviour in time is given by the probability density function (pdf) that the system makes a transition at a given time which leads it into a new configuration.

The configurations of the system (also termed 'states', in the following) can be numbered by an index that orders all the possible combinations of all the states of the components of the system. More specifically, let $k_n \in Z$ denote the index that identifies the configuration reached by the plant at the n-th transition and t_n be the time at which the transition has occurred.

Consider the generic transition which has occurred at time t' with the system entering state k'. The probabilities which govern the occurrence of the next system transition at time t which lead the system into state k are (Fig. 2.1):

- $T(t|t',k')dt$ = conditional probability that the system makes the next transition between t and $t+dt$, given that the previous transition has occurred at time t' and that the system had entered in state k'.
- $C(k|k',t)$ = conditional probability that the system enters in state k by effect of the transition occurring at time t with the system originally in state k'.

Fig. 2.1 *Transition (t',k') ⇒ (t,k)*

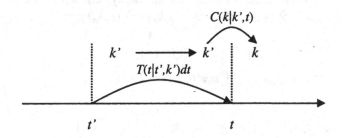

The probabilities above defined are normalized as follows:

$$\int_{t'}^{\infty} T\left(t|t',k'\right)dt \leq 1 \tag{2.1}$$

$$\sum_{k\in\Omega} C\left(k|k',t\right)=1; \quad C\left(k|k,t\right)=0 \tag{2.2}$$

where Ω is the set of all possible states of the system. In the following, unless otherwise stated, all the summations on the system states will be performed on all the values $k\in\Omega$. Note that $T(t|t', k')$ may not be normalized to one since with probability $1- \int T(t|t',k')dt$ the system may fall at t' in a state k' from which it cannot exit, called absorbing state.

The two probability functions introduced form the so called *probabilistic transport kernel* for the transition (t',k') → (t,k) :

$$K\left(t,k|t',k'\right)=T\left(t|t',k'\right)C\left(k|k',t\right) \tag{2.3}$$

As mentioned above, the system stochastic behaviour in time is fully captured by the probability density function that the system makes a transition at a time between t and $t+dt$ and enters state k as a result of the transition. This pdf, denoted as $\psi(t,k)$, can be written as the series of the partial probability densities $\psi^n(t,k)$, i.e. the pdfs that the system enters the n-th transition at time t which leads it into state k.

Let us suppose that at the initial time t^*, the system is in state k^*. The subsequent partial densities representing the system stochastic evolution $(t^*,k^*) \rightarrow (t_1,k_1) \rightarrow (t_2,k_2) \rightarrow \ldots \rightarrow (t_{n-1},k_{n-1}) \rightarrow (t_n,k_n)$ are:

$$\psi^1(t_1,k_1) = K(t_1,k_1|t^*,k^*)$$

$$\psi^2(t_2,k_2) = \sum_{k_1} \int_{t^*}^{t_2} dt_1 \psi^1(t_1,k_1) K(t_2,k_2|t_1,k_1)$$

$$\ldots$$

$$\psi^n(t_n,k_n) = \sum_{k_{n-1}} \int_{t^*}^{t_n} dt_{n-1} \psi^{n-1}(t_{n-1},k_{n-1}) K(t_n,k_n|t_{n-1},k_{n-1}) \qquad (2.4)$$

Substituting t_n, k_n by t, k and t_{n-1}, k_{n-1} by t' and k', the expression for ψ^n at the generic time t and state k becomes:

$$\psi^n(t,k) = \sum_{k'} \int_{t^*}^{t} dt' \psi^{n-1}(t',k') K(t,k|t',k') \qquad (2.5)$$

The probability density $\psi(t,k)$ of a transition at time t as a result of which the system enters state k is given by the series of all the possible partial transition densities, viz.

$$\psi(t,k) = \sum_{n=0}^{\infty} \psi^n(t,k) = \psi^0(t,k) + \sum_{k'} \int_{t^*}^{t} dt' \psi(t',k') K(t,k|t',k') \qquad (2.6)$$

This is the Boltzmann integral equation for the probability density of a system transition to state k occurring at time t. The solution to this equation can be obtained by standard analytical or numerical methods, under given simplifying assumptions, or with the Monte Carlo method under realistic assumptions. In this latter case, a large number $M \gg 1$ of system life *histories* is performed, each of which is a realization, called random walk, of the system stochastic process of evolution in time. The ensemble of the random walks realized by the Monte Carlo histories allows estimating the transition probability density $\psi(t,k)$.

To generate the random walks, the partial transition densities are first re-written as follows, by successive substitutions:

$$\psi^1(t_1,k_1) = K(t_1,k_1|t^*,k^*)$$

$$\psi^2(t_2,k_2) = \sum_{k_1} \int_{t^*}^{t_2} dt_1 \psi^1(t_1,k_1) K(t_2,k_2|t_1,k_1) = \sum_{k_1} \int_{t^*}^{t_2} dt_1 K(t_1,k_1|t^*,k^*) K(t_2,k_2|t_1,k_1)$$

$$\psi^3(t_3,k_3) = \sum_{k_2} \int_{t^*}^{t_3} dt_2 \psi^2(t_2,k_2) K(t_3,k_3|t_2,k_2)$$

(2.7)

$$= \sum_{k_1,k_2} \int_{t^*}^{t_3} dt_2 \int_{t^*}^{t_2} dt_1 K(t_1,k_1|t^*,k^*) K(t_2,k_2|t_1,k_1) K(t_3,k_3|t_2,k_2)$$

...

$$\psi^n(t,k) = \sum_{k_1,k_2,...,k_{n-1}} \int_{t^*}^{t} dt_{n-1} \int_{t^*}^{t_{n-1}} dt_{n-2} ... \int_{t^*}^{t_2} dt_1 \cdot K(t_1,k_1|t^*,k^*) K(t_2,k_2|t_1,k_1) ... K(t,k|t_{n-1},k_{n-1})$$

Eq. (2.7) shows that $\psi^n(t,k)$ can be calculated from the integration of the transport kernels over all possible random walks constituted of n intermediate transitions:

$$(t^*,k^*) \rightarrow (t_1,k_1) \rightarrow (t_2,k_2) \rightarrow ... \rightarrow (t_{n-1},k_{n-1}) \rightarrow (t,k) \qquad (2.8)$$

Conceptually, the Monte Carlo estimate of the multi-dimensional integral defining $\psi^n(t,k)$ amounts to simulating a large number M of random walks and counting the number of these random walks whose n-th transition indeed occurs at time t, with the system indeed entering the state k: the frequency of these occurrences, i.e. the ratio of the number of occurrences over M, gives an estimate of $\psi^n(t,k)$. Actually, if during the simulation of the random walks the occurrence of a transition at time t which leads the system into state k is recorded independently of the order of the transition, then the frequency of such occurrences over the M simulated histories gives directly an estimate of $\psi(t,k)$.

From the point of view of the implementation into a computer code, the system mission time is subdivided in N_t intervals of length Δt and to each time interval N counters are associated, one for each state of the system. Every time the system undergoes a transition of any order $n=1,2,...$, for which the system enters state k at time $t \in \Delta t_l$, the content C_{lk} of the lk-th counter is incremented by 1. At the end of all the M Monte Carlo histories, C_{lk} is equal to the number of occurrences of the system undergoing a transition (of any order) at time $t \in \Delta t_l$ which leads it into state k. An estimate of the transition probability density is then:

$$\psi(t,k) \approx \frac{1}{\Delta t} \int_{\Delta t_l} \psi(\tau,k) d\tau \approx \frac{1}{\Delta t} \frac{C_{lk}}{M} \quad t \in \Delta t_l; l=1,2,...,N_t; k=1,2,...,N \qquad (2.9)$$

2.3   Monte Carlo simulation for system unreliability and unavailability estimation

Let us consider a system functioning at time $t=0$ and let Γ be the subset of the system fault states. The unreliability $F(t)$ of the system at time t is the probability that a system failure occurs before time t, i.e. that the system enters one of its fault states prior to t. This probability is given by the sum of the probabilities of all mutually exclusive random walks which prior to t lead the system into a fault state $k \in \Gamma$, i.e.

$$F(t) = \sum_{k \in \Gamma} \int_0^t \psi(\tau, k) d\tau \qquad (2.10)$$

where $\psi(\tau, k)$ is the probability density of undergoing a transition (of any order) at time τ which leads the system into state k, i.e. of entering state k at time τ. To estimate $F(t)$ by Monte Carlo simulation, M random walks are sampled from the transport kernels (Fig. 2.2). For each random walk, in which the system enters a state $k \in \Gamma$ at time $\tau < t$, a 1 is cumulated in the counter associated to the time t: the estimate $\hat{F}(t)$ of the unreliability at time t is given by the cumulative value in the counter at the end of the M simulated random walks divided by M.

Fig. 2.2 MC estimation procedure of unreliability $F(t)$. In the first history, the system enters the failed configuration $k_1 \in \Gamma$ at $\tau_{in}^1 < t$ and exits at τ_{out}^1; then it enters another, possibly different, failed configuration $k_2 \in \Gamma$ at $\tau_{in}^2 > t$ and does not exit before the mission time T_M. However, for the unreliability estimation of $F(t)$, with $t < \tau_{in}^2$, this latter part of the random walk does not matter because the system has already passed the time of interest t, before which eventual failure occurrences are recorded in the cumulative counter $C^R(t)$. In the second history, no transitions at all occur before t. Hence, the counter $C^R(t)$ remains unmodified. In the third history, the system enters the only failed configuration $k_3 \in \Gamma$, possibly different from the previous ones, at $\tau_{in}^3 > t$ and exits at τ_{out}^3; since the only system failure in the history occurs after t, the counter $C^R(t)$ remains unmodified. In the end, the quantity $\hat{F}(t)$, frequency of failure occurrences before t, gives the MC estimate of the system unreliability at time t, $F(t)$.

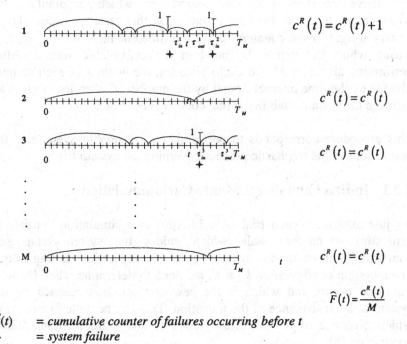

$$c^R(t) = c^R(t) + 1$$

$$c^R(t) = c^R(t)$$

$$c^R(t) = c^R(t)$$

$$c^R(t) = c^R(t)$$

$$\hat{F}(t) = \frac{c^R(t)}{M}$$

$C^R(t)$ = *cumulative counter of failures occurring before t*

✚   = *system failure*

⌒   = *system evolution in time from one transition to the next*

Similarly, the system unavailability at time t, $q(t)$, i.e. the probability that the system is in a fault state at time t, is given by the sum of the probabilities of all mutually exclusive random walks which lead the system in a state $k \in \Gamma$ at a time τ prior to t and in which the system remains beyond time t, i.e.

$$q(t) = \sum_{k \in \Gamma} \int_0^t \psi(\tau, k) R_k(\tau, t) d\tau \qquad (2.11)$$

where $R_k(\tau, t)$ is the probability of the system not exiting before t from the failed state k entered at $\tau < t$.

An estimate of $q(t)$ may be obtained by MC simulation of the system random walks as was done for the unreliability estimation. Let us consider a single random walk and suppose that the system enters a failed state $k \in \Gamma$ at time τ_{in}, exiting from it at the next transition at time τ_{out}; as before, the time is suitably discretized in intervals of length Δt and cumulative counters $C^A(t)$ are introduced which accumulate the contributions to $q(t)$ in the time channels: in this case, we accumulate a unitary weight in the counters for all the time channels within $[\tau_{in}, \tau_{out}]$ during which the system is found in the unavailable state k. After performing all the M Monte Carlo histories, the content of each counter divided by the time interval Δt and by the number of histories M gives an estimate of the unavailability at that counter time (Fig. 2.3).

This procedure corresponds to performing an ensemble average of the realizations of the stochastic process governing the system life.

2.3.1 Indirect and direct Monte Carlo simulation

As just explained, each trial of a Monte Carlo simulation consists in generating a random walk which guides the system from one configuration to another, at different times. During a trial, starting from a given system configuration k' at t', we need to determine when the next transition occurs and which is the new configuration reached by the system as a consequence of the transition. This can be done in two ways which give rise to the so-called "indirect" and "direct" Monte Carlo approaches [8].

The indirect approach consists in sampling first the time t of a system transition from the corresponding conditional probability density

$T(t|t',k')$ of the system performing at time t one of its possible transitions out of k' entered at the previous transition at time t'. Then, the transition to the new configuration k actually occurring is sampled from the conditional probability $C(k|t,k')$ that the system enters the new state k given that a transition has occurred at t starting from the system in state k'. The procedure then repeats to the next transition.

The direct Monte Carlo simulation method differs from the previous one in that the system transitions are not sampled by considering the kernel distributions for the whole system but rather by sampling directly the times of all possible transitions of all individual components of the system and then arranging the transitions along a timeline in increasing order, in accordance to their times of occurrence. The component which actually performs the transition is the one corresponding to the first transition in the timeline. Obviously, this timeline is updated after each transition occurs, to include the new possible transitions that the transient component can perform from its new state. In other words, during a history starting from a given system configuration k' at t', we sample the times of transition $t^i_{j'_i \to m_i}, m_i = 1,2,...,N_{S_i}$, of each component i, $i=1,2,...,N_c$ leaving its current state j'_i and arriving to the state m_i from the corresponding transition time probability distributions $f_T^{i,j'_i \to m_i}(t|t')$. The time instants $t^i_{j'_i \to m_i}$ thereby obtained are then arranged in ascending order along a timeline from $t_{\min}$ to $t_{\max} \leq T_M$. The clock time of the trial is then moved to the first occurring transition time $t_{\min}= t^*$ in correspondence of which the system configuration is changed, i.e. the component i^* undergoing the transition is moved to its new state m_i^*. At this point, the new times of transition $t^{i^*}_{m_i^* \to l_i^*}, l_i^* = 1,2,...,N_{S_{i^*}}$, of component i^* out of its current state m_i^* are sampled from the corresponding transition time probability distributions, $f_T^{i^*,m_i^* \to l_i^*}(t|t^*)$, and placed in the proper position of the timeline. The clock time and the system are then moved to the next first occurring transition time and corresponding new configuration, respectively, and the procedure repeats until the next first occurring transition time falls beyond the mission time, i.e. $t_{\min} > T_M$.

Compared to the previous indirect method, the direct approach is more suitable for systems whose components' failure and repair behaviours are represented by different stochastic distribution laws.

Fig. 2.3 MC estimation procedure of unavailability $q(t)$. In the first history, the system enters the failed configuration at $\tau_{in}^1 < t$ and exits at τ_{out}^1; then it enters another, possibly different, failed configuration $k_2 \in \Gamma$ at $\tau_{in}^2 > t$ and does not exit before the mission time T_M. However, for the unavailability estimation of $q(t)$, with $t < \tau_{in}^2$, this latter part of the random walk does not matter because the system has already passed the time of interest t, before which eventual failure occurrences are recorded in the cumulative counter $C^A(t)$. In the second history, no transitions at all occur before t. Hence, the counter $C^A(t)$ remains unmodified. In the third history, the system enters the only failed configuration $k_3 \in \Gamma$, possibly different from the previous ones, at $\tau_{in}^3 > t$ and exits at τ_{out}^3; since the only system failure in the history occurs after t, the counter $C^A(t)$ remains unmodified. In the end, the quantity $\hat{q}(t)$, gives the MC estimate of the system unreliability at time t, $q(t)$.

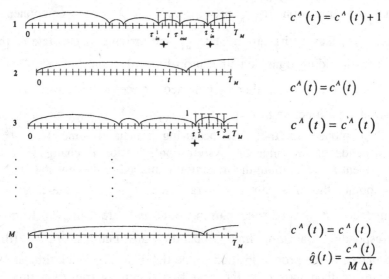

$$c^A(t) = c^A(t) + 1$$

$$c^A(t) = c^A(t)$$

$$c^A(t) = c^A(t)$$

$$c^A(t) = c^A(t)$$

$$\hat{q}(t) = \frac{c^A(t)}{M \Delta t}$$

$c^A(t)$ = *cumulative counter which accumulates the contributions to $q(t)$ in the time channels .*

✛   = *system failure*

⌢   = *system evolution in time from one transition to the next*

2.4 References

[1] Cashwell, E. D., Everett, C. J., *A Practical Manual on the Monte Carlo Method for Random Walk Problems,* Pergamon Press, N. Y., 1959.

[2] Rubinstein, R. Y., *Simulation and the Monte Carlo Method,* Wiley, 1981.

[3] Kalos, M. H., Whitlock, P. A., *Monte Carlo Methods. Volume 1: Basics,* Wiley, 1986.

[4] Lux, I., Koblinger, L., *Monte Carlo Particle Transport Methods: Neutron and Photon Calculations,* CRC Press, 1991.

[5] Henley, E. J., Kumamoto, H., *Probabilistic Risk Assessment,* IEEE Press, 1991.

[6] Dubi, A., *Monte Carlo Applications in Systems Engineering,* Wiley, 1999.

[7] Marseguerra, M., Zio, E., *Basics of the Monte Carlo Method with Application to System Reliability,* LiLoLe- Verlag GmbH (Pbl. Co. Ltd.), 2002.

[8] Labeau, P.E., Zio, E., *Procedures of Monte Carlo Transport Simulation for Applications in System Engineering,* Reliability Engineering and System Safety,**77** , 2002, pp 217-228.

3. MARKOV CHAIN MONTE CARLO FOR APPLICATIONS TO RELIABILITY AND AVAILABILITY ANALYSIS

3.1 Introduction

In general terms, Markov Chain Monte Carlo (MCMC) algorithms [1-4] offer an effective way for sampling from complicated probability distributions in high-dimensional spaces. This is useful in such tasks as image reconstruction, parameter identification, computing the equilibrium distribution and associated energy levels of statistical mechanics systems, inverse problem solving and more generally Bayesian posterior inference. Detailed reviews on the subjects may be found e.g. in [5-11].

Consider a vector of random variables ϑ, its (possibly not normalized) probability density function $\pi(\vartheta)$ on some space Ω and the expectation of a function $f(\vartheta)$,

$$E[f(\vartheta)] = \int_{\Omega} f(\vartheta)\pi(\vartheta)d\vartheta \qquad (3.1)$$

When Ω is high-dimensional and/or $\pi(\vartheta)$ is a complicated function, the computation of $E[f(\vartheta)]$ by direct integration (either analytic or numerical) is in general infeasible [8]. One can then resort to Monte Carlo simulation (chapter 2) [12], which amounts to sampling n values $\vartheta_1, \vartheta_2, ..., \vartheta_n$ from $\pi(\vartheta)$, and then estimating the expectation $E[f(\vartheta)]$ as

$$E[f(\vartheta)] \approx \frac{1}{n}\sum_{i=1}^{n} f(\vartheta_i) \qquad (3.2)$$

This gives an unbiased estimate, with a standard deviation of order $\sqrt{n}$ [5,8,10,12]. However, when the functional form of $\pi(\vartheta)$ is complicated, sampling from this distribution becomes extremely difficult.

The MCMC algorithms provide an alternative way for sampling from $\pi(\vartheta)$. This approach is based on the construction of a *Markov chain* (chapter 1) in the space Ω, whose stationary distribution is the target function $\pi(\vartheta)$ [5,8].

While MCMC algorithms are used in many fields of science (such as statistical mechanics and computer science), their most widespread application is in Bayesian statistical inference [10,13-15], in which case the target distribution is the posterior distribution $\pi(\vartheta \mid x)$, given a set of measures x. The variable ϑ may represent, for example, the unknown parameters of a (known) model, which we would like to estimate on the basis of the experimentally accessible data x. If the Markov chain is run for long enough, the distribution of the generated values $\vartheta_i \mid x$, $i=1,2,...,n$ becomes stationary and approaches $\pi(\vartheta \mid x)$, independently of the starting point of the chain [5,8].

MCMC is, of course, not the only way to sample from complicated probability distributions. Other possible sampling algorithms, not considered here, include *rejection sampling* and *importance sampling* [12].

3.2 The Metropolis-Hastings algorithm

We begin by illustrating the Metropolis-Hastings algorithm, which represents one of the most popular means of practically implementing MCMC for Bayesian inference [1,2].

According to Bayes' theorem, the general structure of $\pi(\vartheta\,|\,x)$ for a given dataset x is

$$\pi(\vartheta\,|\,x) \propto L(x\,|\,\vartheta)p(\vartheta) \tag{3.3}$$

where $L(x\,|\,\vartheta)$ is the likelihood of the dataset given the parameters, which depends on the functional form of the chosen model, and $p(\vartheta)$ is the prior distribution of the parameters, which represents the *a priori* knowledge on their values. Both $L(.\,|\,.)$ and $p(.)$ are assumed to be known functions. The proportionality sign means that the target distribution $\pi(\vartheta\,|\,x)$ is known up to its normalization factor.

The Markov chain is initialized with a randomly chosen value ϑ^0. The sequence $\vartheta^1, \vartheta^2, ..., \vartheta^n$ of the chain is determined by resorting to a suitable *kinetics* $K(\vartheta'\,|\,\vartheta)$ which allows generating a new proposal value ϑ' from the previous. A convenient choice, often adopted in practice, is to take symmetric kinetics $K(\vartheta'\,|\,\vartheta)$, so that $K(\vartheta'\,|\,\vartheta) = K(\vartheta\,|\,\vartheta')$. This was indeed the choice in the original algorithm by Metropolis et al. [1]; the generalization to non-symmetric kinetics is due to Hastings [2].

Given ϑ^j at the generic j^{th} iteration of the chain, the proposal value ϑ' is accepted to become the new value ϑ^{j+1} in the chain with probability

$$\alpha = \min\left\{1, \frac{\pi(\vartheta'\,|\,x)}{\pi(\vartheta^j\,|\,x)} \frac{K(\vartheta^j\,|\,\vartheta')}{K(\vartheta'\,|\,\vartheta^j)}\right\} \tag{3.4}$$

Otherwise, the proposal value ϑ' is rejected and the new value ϑ^{j+1} is taken equal to the previous one, i.e. $\vartheta^{j+1} = \vartheta^j$. Note that the functions $\pi(\vartheta'|x)$ and $\pi(\vartheta|x)$ enter the acceptance probability (3.4) only as a ratio; therefore, the distribution $\pi(.|x)$ needs only to be known up to a normalization factor. This is undoubtedly a major benefit of the algorithm, since for Bayesian inference problems the computation of the normalization factor is usually extremely cumbersome and must be performed by numerical integration [5,8,10]. Moreover, when $K(.|.)$ is a symmetric function the ratio simplifies and depends only on the target distribution $\pi(.|x)$.

The initial m values of the chain (the so-called *burn-in period*) are discarded and the remaining $n-m$ values, under mild assumptions on the assigned kinetics $K(.|.)$ [1,2,5-10,13], approximately obey to the target distribution $\pi(\vartheta|x)$. The approximation is due to the fact that the values ϑ^j of the chain are by definition correlated [1,2].

3.2.1 Application to the estimation of the failure rate of a deteriorating component

In this Section, the application of the MCMC Metropolis-Hastings algorithm to an example of Bayesian inference is illustrated. Consider a component with constant failure rate λ. On the basis of a sample of collected inter-failure times $x = [t_1, t_2, ..., t_N]$, one would like to estimate the (unknown) parameter λ.

Within a Bayesian inference framework, the target posterior distribution is $\pi(\lambda|x) \propto L(x|\lambda)p(\lambda)$, where $L(x|\lambda)$ is the likelihood function and $p(\lambda)$ is the prior distribution of the unknown parameter. Concerning $L(x|\lambda)$, its functional form is

$$L(\lambda|x) = \lambda^N e^{-\lambda \sum_{i=1}^{N} t_i} \tag{3.5}$$

The prior distribution $p(\lambda)$ is chosen uniform on the range $[0, \lambda_{max}]$, where λ_{max} is conveniently fixed much larger than the expected value of λ.

Finally, as symmetric kinetics ($K(\lambda'|\lambda) = K(\lambda|\lambda')$) a Gaussian pdf $K(a|b)$ is chosen, with mean b and standard deviation to be adjusted so to achieve good mixing properties.

The chain is initialized with a given value λ^0; given the value λ^j of the parameter at the generic j^{th} iteration, a proposal value λ' for the parameter value at the next iteration $j+1$ is sampled from the kinetics and accepted with probability

$$\alpha = \min\left\{1, \frac{L(x|\lambda')}{L(x|\lambda^j)}\right\} \tag{3.6}$$

by virtue of the fact that the kinetics is assumedly symmetric and the uniform distributions $p(\lambda')$ and $p(\lambda)$ simplify in the ratio. Practically, a uniform random number r is drawn in the range $[0,1)$: if $r \leq \alpha$, the proposal value is accepted and $\lambda^{j+1} = \lambda'$; otherwise, the value is rejected and the parameter value remains unchanged, i.e. $\lambda^{j+1} = \lambda^j$.

The large sample of parameter values generated in the iterative chain approximately obeys to the target distribution $\pi(\lambda|x) \propto L(x|\lambda)p(\lambda)$ (once the burn-in period, i.e. approximately the first 20% of the chain, has been discarded) and can be used to perform Bayesian inference on the observed component subject to failures.

Fig. 3.1 the convergence of the chain is shown as a function of the number of iterations. For this numerical example, the standard deviation of the kinetics has been set to $\sigma = 10^{-4}$, the number of collected samples has been set to $N = 100$ and the algorithm has been run for 10^5 iterations. The true failure rate λ used to generate the dataset was $\lambda = 10^{-2}$. The convergence is rapidly attained after one or two thousand steps.

Fig. 3.1 The convergence of the Markov Chain as a function of the number of iterations

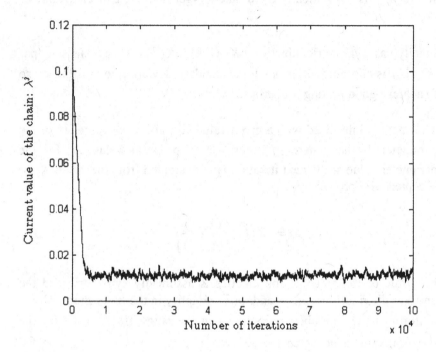

In Fig. 3.2 the posterior distribution $\pi(\lambda \mid x)$ of the failure rate is shown. The burn-in period has been set to $m = 2 \cdot 10^4$ iterations. The distribution reveals a good agreement with the true failure rate, which is plotted as a dashed line. The accuracy of the estimate could be further improved by resorting to a Bayesian updating strategy, as new information becomes available. This issue is discussed later in Section 3.5.

Fig. 3.2 Posterior distribution $\pi(\lambda \mid x)$ **of the failure rate. A good agreement is found with the true parameter** $\lambda = 10^{-2}$ **used to generate the data (plotted as a dashed line)**

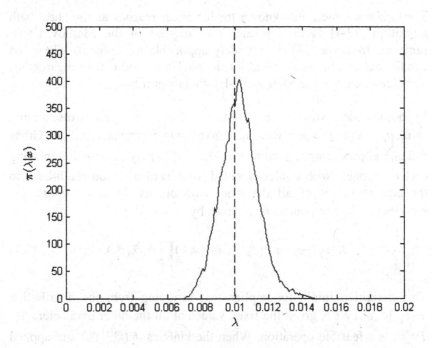

3.3   The Gibbs sampler

The Gibbs sampler, also known for historical reasons as the "heat-bath algorithm" [3-4] is conceptually the simplest of the Markov Chain sampling methods [14]. It is widely applicable to problems where the conditional distributions are easily sampled from and can be regarded as a particular case of the Metropolis-Hastings algorithm.

Suppose one wishes to sample from a joint distribution $\pi(\vartheta_1, \vartheta_2, ..., \vartheta_n \mid x)$, conditional to a given set of measures x. The Gibbs sampler allows sampling from $\pi(\vartheta_1, \vartheta_2, ..., \vartheta_n \mid x)$ by repeatedly replacing each component with a value sampled from its distribution conditional to the current values of all the other components. In other words, the kinetics for the component ϑ_k is given by

$$K(\vartheta_k ' \mid \vartheta_k) = P(\vartheta_k ' \mid x, \vartheta_i, i \neq k) \prod_{i \neq k} \delta(\vartheta_i, \vartheta_i ') \qquad (3.7)$$

This definition implicitly implies that sampling from the distribution $P(\vartheta_k ' \mid x, \vartheta_i, i \neq k)$, given the fixed values of all the other parameters ϑ_i, $i \neq k$, is a feasible operation. When the kinetics $K(\vartheta_k ' \mid \vartheta_k)$ are applied in sequence to all the parameters ϑ_k, the algorithm can be described as a Markov Chain with transition probability

$$T = \prod_k K(\vartheta_k ' \mid \vartheta_k) \qquad (3.8)$$

The procedure for generating at the iteration $j+1$ the new parameters vector $[\vartheta_1, \vartheta_2, ..., \vartheta_n]^{j+1}$ given their values $[\vartheta_1, \vartheta_2, ..., \vartheta_n]^j$ computed at the previous step j of the chain is the following, on a component-by-component basis:

 − sample ϑ_1^{j+1} from the distribution of ϑ_1 given $\vartheta_2^j, \vartheta_3^j, ..., \vartheta_n^j$ and x. Update $\vartheta_1^j \rightarrow \vartheta_1^{j+1}$;

−sample ϑ_2^{j+1} from the distribution of ϑ_2 given $\vartheta_1^{j+1}, \vartheta_3^j, ..., \vartheta_n^j$ and x. Update $\vartheta_2^j \to \vartheta_2^{j+1}$;

−...

−sample ϑ_n^{j+1} from the distribution of ϑ_n given $\vartheta_1^{j+1}, \vartheta_2^{j+1}, ..., \vartheta_{n-1}^{j+1}$ and x. Update $\vartheta_n^j \to \vartheta_n^{j+1}$;

Technically speaking, the Gibbs sampler is based on a concatenation of n Metropolis-Hastings algorithms, one for each component ϑ_k of the parameters vector. In this concatenation, there are practically n target distributions (which are often called "full conditional distributions"), with the proposal and target distributions having the same form. Due to this property, the acceptance probability of each proposed value is always one: in other words, the candidate component ϑ_k' in the chain is always accepted. A distinct merit of the algorithm is that the full conditional distribution is univariate, so that sampling from it should be easier [5].

Under suitable conditions [5,14], the Markov Chain thus generated will converge towards the target equilibrium distribution $\pi(\vartheta_1, \vartheta_2, ..., \vartheta_n \,|\, x)$ and the produced sample can be used for performing Bayesian inference as shown in the previous Section. The algorithm crucially depends on the possibility of (efficiently and quickly) performing the conditional sampling illustrated above: it must be remarked that not all problems can be suitably tackled using the Gibbs sampler, since the conditional distribution might be highly complex or even analytically intractable. In these cases, the problem can usually be tackled with the Metropolis-Hastings algorithm. Generally speaking, the Gibbs sampler may be particularly appropriate for systems involving discrete variables defined on a finite set or for continuous variables whose conditional distributions have standard forms for which efficient sampling techniques have been developed [14].

3.3.1 Application to the estimation of the parameters of a rare failures process

In this Section, the Gibbs sampler is applied to the Bayesian estimation of the unknown model parameters for a component whose number of failures in a given time period is suitably described by a Poisson distribution with parameter λ and an *ad hoc* "dispersed hyper-parameter" b which carries little information on the value λ and is introduced for illustrative purposes. The reader is referred to References [4-5,10] for further details on this issue.

Given a sample set $x = [k_1, k_2, ..., k_N]$ of the number of failures k_i, $i = 1, 2, ..., N$, occurred in the N observed time intervals of component life, the target distribution is the posterior $\pi(\vartheta | x)$ of the model parameters $\vartheta \equiv (a, b)$.

For this problem, it is assumed that the posterior distribution has the following "hierarchical" structure, which is often encountered in practice [5,10]:

$$\pi(\lambda, b \,|\, x) \propto L(x \,|\, \lambda) p(\lambda \,|\, b) p(b) \qquad (3.9)$$

where $L(x \,|\, \lambda) = \prod_{i=1}^{N} \dfrac{\lambda^{k_i} e^{-\lambda}}{k_i!}$ is the likelihood function based on the underlying Poisson distribution, $p(\lambda \,|\, b) = Gamma(0.5, b) = \dfrac{1}{\Gamma(0.5)b^{0.5}} \lambda^{-0.5} e^{-\lambda/b}$ is a Gamma distribution and finally $p(b) = IGamma(0,1) = \dfrac{e^{-1/b}}{b}$ is an Inverse Gamma distribution.

The hierarchical structure of the posterior distribution is particularly suitable to adopt the Gibbs sampler. The algorithm will proceed as follows:

i. choose at random the starting pair of parameter values for the Markov Chain: $[\lambda, b]^0$

ii. at the generic j^{th} iteration, sample a value λ^{j+1} from the full conditional distribution $\pi(\lambda | x, b^j)$, using the most up-to-date information (the parameter b is fixed at the value b^j)

iii. then, sample a value b^{j+1} from the full conditional distribution $\pi(b | x, \lambda^{j+1})$ using the most up-to-date information (the parameter λ is fixed at the value λ^{j+1})

iv. update the Markov Chain with the new pair of values $[\lambda, b]^{j+1}$

In the case under analysis, it is easy to see that the full conditional distribution $\pi(\lambda | x, b)$ when b is frozen reads

$$\pi(\lambda | x, b) \propto \lambda^{\sum_{i=1}^{N} k_i - 0.5} e^{-\lambda(N + 1/b)} \propto Gamma\left(\sum_{i=1}^{N} k_i + 0.5, \frac{b}{Nb+1}\right),$$ whereas the

full conditional distribution $\pi(b | x, \lambda)$ when λ is fixed reads

$$\pi(b | x, \lambda) \propto \frac{1}{b^{0.5}} e^{-\lambda/b} \frac{e^{-b}}{b} \propto IGamma\left(0.5, \frac{1}{\lambda+1}\right).$$ The Gamma and

Inverse Gamma are straightforwardly sampled from, e.g. with standard Matlab routines.

Fig. 3.3 displays the posterior distribution $\pi(\lambda | x)$ of the parameter λ. A good agreement is found with the true parameter $\lambda = 1$ used to generate the data (plotted as a dashed line)

Fig. 3.3 Posterior distribution $\pi(\lambda\,|\,x)$ **of the parameter** λ. **A good agreement is found with the true parameter** $\lambda = 1$ **used to generate the data (plotted as a dashed line)**

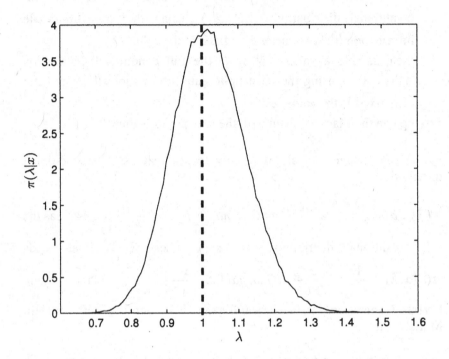

3.4 The reversible-jump MCMC algorithm

An important class of problems in Bayesian inference concerns the detection of *changepoints*, i.e. instances at which the underlying model parameters have a (step) change. The aim of the MCMC algorithm, in this case, would be to estimate the location(s) of the changepoint(s) as well as the values of the parameters before and after the transition(s), on the basis of the observed dataset. This problem has been extensively studied in the framework of Bayesian posterior analysis (see e.g. [16-21] and References therein).

Application of the standard Metropolis-Hastings algorithm, however, would entail preliminarily fixing the number k of changepoints which are likely to be present in the observed data. In many applications, though, either the information concerning the number of changepoints is not available, or it is desirable to estimate also the uncertainty on the number of such changepoints. Thus, the dimensionality of the parameter space is itself a stochastic variable, with posterior (discrete) distribution $p(k \mid x)$. An efficient solution to perform Bayesian analysis in the context of an unknown number of changepoints has been suggested by Green in a seminal paper [17] and is known as *reversible-jump* Markov chain Monte Carlo algorithm.

Suppose that there are k changepoints at locations $s_0 < s_1 < s_2 < ... < s_k < s_{k+1}$ (the values s_0 and s_{k+1} are known and correspond to the first and final observations, respectively). Assume further that the parameter space can be approximated as a stepwise (vector) function ϑ, which assumes the values ϑ_i on each interval $[s_i, s_{i+1})$, with $i = 0,1,...,k$. To keep notation simple, ϑ is assumed to be a scalar, i.e. a single parameter which takes $k+2$ values on the $k+1$ intervals defined by the step locations s_i; extension to the multidimensional case is straightforward.

Let $L(x|\vartheta)$ be the likelihood function of the observed data, $\pi(\vartheta|x) \propto L(x|\vartheta)p(\vartheta)$ the posterior distribution and $p(\vartheta)$ the prior distribution of the parameters.

The prior distribution $p(k)$ for the number of changepoints is assumed to be a Poisson distribution

$$p(k) = \frac{e^{-\lambda}\lambda^k}{k!} \tag{3.10}$$

with parameter λ. The average of the distribution, which coincides with the free parameter λ, is set according to the *a priori* expected number of changepoints. Usually, the pdf $p(k)$ is truncated to some maximal value k_{max}. Once the prior value of k is assigned, a convenient choice for the prior distribution of the step positions s_i is to take the even-numbered order statistics of $2k+1$ points uniformly distributed in the interval $[s_0, s_{k+1}]$. This assumption allows the initial step locations s_i to be more evenly spaced on $[s_0, s_{k+1}]$. Finally, the prior distribution for the parameter ϑ is usually assumed to be *uninformative* and a convenient choice is a Gamma pdf of the kind

$$p(\vartheta) = \frac{\beta^\alpha}{\Gamma(\alpha)}\vartheta^{\alpha-1}e^{-\beta\theta} \tag{3.11}$$

where the parameters α and β are to be assigned. This completes the initial state of the Markov chain.

Four possible random *moves* can be considered:

i.   the parameter value ϑ is varied at a random location s_i;
ii. a randomly chosen step location s_i is moved;
iii. a new step location is created at random in the interval $[s_0, s_{k+1}]$ (*birth move*);
iv. a randomly chosen location s_i is eliminated (*death move*).

These last two moves involve a change in the dimension of the parameter space, so that the standard MCMC theory can not be applied [17].

At each transition, one out of the four possible moves is selected at random, according to some assigned probabilities. Let η_k, χ_k, b_k and d_k be the probabilities of attempting a change in the value of ϑ, a change in the step location, the birth of a new changepoint or the death of an existing changepoint, respectively.

These probabilities are in principle arbitrary: a convenient criterion is to impose their values so that the Markov chain has good *mixing* properties (i.e. fast convergence to the equilibrium target distribution) and the parameter space is thoroughly explored. The normalization is $\eta_k + \chi_k + b_k + d_k = 1$; furthermore, $d_0 = \chi_0 = b_{k_{max}} = 0$. Following [17], η_k is set equal to χ_k, when $k \neq 0$, and the probabilities of a birth or death event are chosen of the following form:

$$b_k = c\min\left\{1, \frac{p(k+1)}{p(k)}\right\} \text{ and } d_k = c\min\left\{1, \frac{p(k-1)}{p(k)}\right\} \qquad (3.12)$$

respectively. This choice ensures the equilibrium probability balance $b_k p(k) = d_{k+1} p(k+1)$ between birth and death moves, for any k. The constant c can be set so to achieve the good mixing properties of the algorithm; it turns out that an optimal choice is to assume c as large as possible, subject to the constraint that $b_k + d_k \leq 0.9$, for any k. The four different moves are generated as follows:

• When a change in the parameter values is sampled (with probability η_k), a variation in ϑ is obtained as follows. First, a value ϑ_i is chosen at random; then, a new value ϑ_i' is proposed such that $\log(\frac{\vartheta_i'}{\vartheta_i})$ is uniformly distributed in the range $[-0.5, 0.5]$. The proposed value ϑ_i' is accepted with probability

$$\alpha_\eta = \min\left\{1, \ell\left(\frac{\vartheta_i{}'}{\vartheta_i}\right)^\alpha e^{-\beta(\vartheta_i{}'-\vartheta_i)}\right\}$$ (3.13)

where $\ell = \dfrac{L(x|\vartheta_i{}')}{L(x|\vartheta_i)}$ is the likelihood ratio. Given the piecewise structure

of the parameter function ϑ, the likelihood ratio assumes a simple form, since all the contributions cancel out, but for those related to the intervals enclosing s_i.

• A step position move is sampled as follows (with probability χ_k): first, a location s_i is chosen at random among the existing values; then, a proposal value $s_i{}'$ is drawn uniformly in the range $[s_{i-1}, s_{i+1}]$. The acceptance probability is

$$\alpha_\chi = \min\left\{1, \ell\frac{(s_{i+1}-s_i{}')(s_i{}'-s_{i-1})}{(s_{i+1}-s_i)(s_i-s_{i-1})}\right\}$$ (3.14)

• The birth move is based on the *dimension-matching* criterion introduced in [17], which enables to switch from a parameter space of dimension k to another space of dimension $k+1$. The interested reader can refer to the original reference for the details. First, a new location s^* is chosen at random and the interval (s_i, s_{i+1}) which surrounds s^* is determined; then, two new values $\vartheta_i{}'$ and $\vartheta_{i+1}{}'$ are proposed on the two intervals (s_i, s^*) and (s^*, s_{i+1}), respectively. These new values are conceived as a *perturbation* of the original value ϑ_i on the interval (s_i, s_{i+1}); the perturbation is introduced by imposing

$$\frac{\vartheta_{i+1}{}'}{\vartheta_i{}'} = \frac{1-r}{r}$$ (3.15)

where r is a random number drawn uniformly in the interval $[0,1)$. The values of the proposal ϑ' are then assigned by imposing e.g. a (geometric or arithmetic) mean structure. Here a simple arithmetic mean structure has been adopted:

$$(s^* - s_i)\vartheta_i' + (s_{i+1} - s^*)\vartheta_{i+1}' = (s_{i+1} - s_i)\vartheta_i \qquad (3.16)$$

The acceptance probability for the birth move can be written in the following form:

$$\alpha_b = \min\{1, \ell PRJ\} \qquad (3.17)$$

where

$$P = \frac{p(k+1)}{p(k)} \frac{2(k+1)(2k+3)}{(s_{k+1} - s_0)^2} \frac{(s^* - s_i)(s_{i+1} - s^*)}{s_{i+1} - s_i} \frac{\beta^\alpha}{\Gamma(\alpha)} \left(\frac{\vartheta_i'\vartheta_{i+1}'}{\vartheta_i}\right)^{\alpha-1} \cdot \\ \cdot e^{-\beta(\vartheta_i' + \vartheta_{i+1}' - \vartheta_i)} \qquad (3.18)$$

is the prior ratio,

$$R = \frac{d_{k+1}(s_{k+1} - s_0)}{b_k(k+1)} \qquad (3.19)$$

is the proposal ratio, and

$$J = \frac{(\vartheta_i' - \vartheta_{i+1}')^2}{\vartheta_i} \qquad (3.20)$$

is the Jacobian of the proposed dimension-changing move $k \to k+1$. If the move is accepted, the locations and the parameters are re-labelled in order to include the newly created proposal values.

- The death move is performed as follows: an existing location s_i is selected at random to be removed; the new parameter on the interval is then determined by reversing the birth move so that the dimension-matching criterion between the $k \to k-1$ and $k \to k+1$ moves is satisfied [17]:

$$(s_i - s_{i-1})\vartheta_i + (s_{i+1} - s_i)\vartheta_i = (s_{i+1} - s_{i-1})\vartheta_{i-1}' \qquad (3.21)$$

The acceptance probability α_d has the same form as the acceptance probability α_b for the birth move, with the due changes in the variables labels [17,20,21]. If the move is accepted, the step locations and the corresponding values of the parameters on the intervals are suitably re-labelled.

The reversible-jump MCMC algorithm proceeds as follows, starting from the initial (prior) values of the Markov chain. A random number is sampled uniformly in $[0,1)$ and a move is selected. The acceptance probability (which depends on the current value of k) is computed. Then, a second uniform random number is drawn and the move is rejected or accepted according to the corresponding acceptance ratio. The values of the chain are updated, a new value of k is determined and the move probabilities are updated in turn. When a large number n of values in the chain have been iteratively sampled, and the burn-in period m (again, usually in the range of 20% of n) has been discarded, the remaining $n - m$ values can be used to perform Bayesian inference as described before.

3.4.1 Application to the estimation of the failure rate of a component subject to degradation or improvement

The above described reversible-jump MCMC algorithm can be applied to the characterization of the time evolution of a component subject to deterioration (or improvement). This well-known problem has been the object of numerous research efforts (see e.g. [20,21-26]).

Consider a system whose initial, constant failure rate is λ_1 ; because of the many human interventions on the component, both during operation and maintenance, and of variations in the operating environment, the reliability of the component is likely to vary during its lifetime. This is mirrored in the experimental dataset $x = \{x_1, x_2, ..., x_N\}$ of the recorded failure times of the component.

If the component deteriorates, due e.g. to aging, the failure times will be closer as time increases [25-26]; on the other hand, if the component experiences a *reliability growth*, as e.g. in the case of prototypes in the testing phase [29-30], the failure times become more separated in time.

For modeling purposes, let us assume that the reliability of the component has only stepwise changes in time. On the basis of the observed set x of failure times, one wishes to determine the k times s_i ($i = 0, 1, ..., k$) at which changes in failure rate occur and the values of the failure rates λ_i on each time interval (s_i, s_{i+1}), between successive changes. Moreover, one also wishes to quantify the uncertainty on the number of changepoints.

Supposing that the underlying stochastic model for the failure times of the component is a time-inhomogeneous Poisson process [17,21,24-26], the log-likelihood function is

$$L(x \mid \lambda) = \sum_{i=1}^{N} \log(\lambda(x_i)) - \int_0^T \lambda(u) du \qquad (3.22)$$

where λ plays the role of the parameter ϑ in the previous notation, $s_0 = 0$ and $s_{k+1} = T$ is the final failure time in the dataset.

The results of the application to a synthetic dataset are displayed in Figs. 3.4 – 3.7. The dataset is composed of 240 failure times of a component whose failure rate changes at $s_1 = 84$ from $\lambda_1 = 1$ to $\lambda_2 = 2\lambda_1$ and at $s_2 = 120$ from λ_2 to $\lambda_3 = 4\lambda_1$, in arbitrary units of inverse of time.

The cumulative number of failures as a function of time is shown in Fig. 3.4. In Fig. 3.5 the posterior probability mass distribution of the number of changepoints k is reported: the most probable value predicted is $k = 2$, but also $k = 3$ and $k = 4$ carry a non-negligible probability.

The conditional probability density functions of the changepoint locations s_i and failure rates λ_i on each subinterval, given a number k of changepoints equal to 2 or 3, are presented in Fig. 3.6 and Fig. 3.7, respectively. Dotted-dashed curves display the probability density functions, given $k = 2$; dotted curves the probability density functions, given $k = 3$. The location of a possible third changepoint (Fig. 3.6, dotted curve centered at $s \approx 50$) has been assigned an almost flat distribution, representative of a strong uncertainty. Correspondingly, two of the four peaks in the probability density functions of the failure rates λ_i conditioned to $k = 3$ (Fig. 3.7, dotted curves centered at $\lambda \approx 1.3$ and $\lambda \approx 1.9$) are very close to each other. In Fig. 3.6 and Fig. 3.7, the probability densities for s_i and λ_i *weighted* on the distribution of k (solid line) are also reported. The distributions are coherent with the values of the parameters used to generate the dataset, as shown inTable 3.1, where the true values of s_i and λ_i are compared to the most probable values (abscissas of the peaks of the corresponding distributions).

Fig. 3.4 Synthetic dataset for the non-homogeneous Poisson model of component failures. Abscissas represent the failure times in arbitrary units, ordinates the cumulative number of failures. Three values of λ have been used to generate the dataset, with $\lambda_3 > \lambda_2 > \lambda_1$

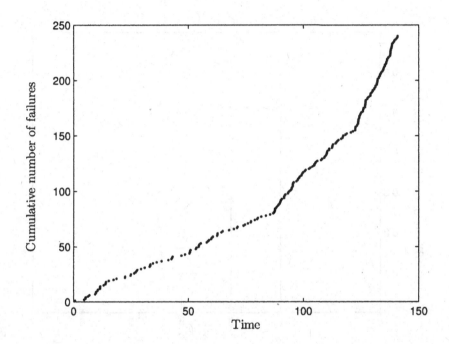

Fig. 3.5 The posterior probability mass distribution of the sampled values k. The reversible-jump MCMC algorithm has correctly identified that the most probable number of changepoints is $k = 2$. The values $k = 3$ and $k = 4$ are also assigned a non-negligible probability

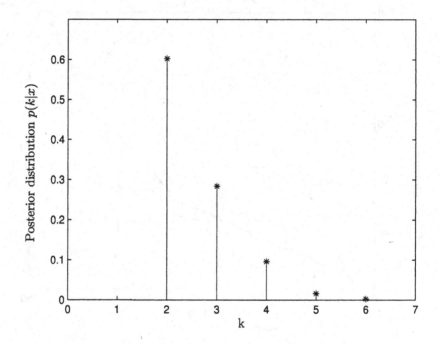

Fig. 3.6 Posterior probability density functions for the locations s_i of the changepoints (expressed in arbitrary units of time), conditioned to the number k of changepoints. Dashed-dotted curves represent the distributions conditional to $k = 2$, whereas dotted curves represent the distributions conditional to $k = 3$. The solid line depicts the probability density of the changepoint locations weighted on the posterior distribution of k. The estimated values are consistent with the parameters used to generate the dataset (Table 3.1)

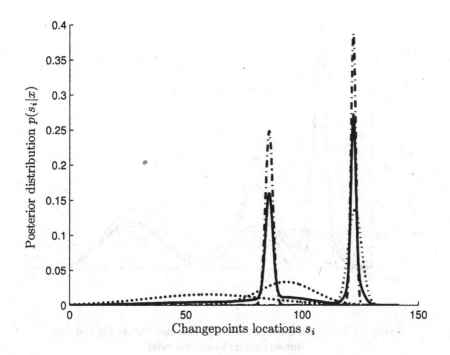

Fig. 3.7 Posterior probability density functions for the failure rates λ_i, conditioned to the number k of changepoints. Dashed-dotted curves represent the distributions conditional to $k = 2$, whereas dotted curves represent the distributions conditional to $k = 3$. The solid line depicts the probability density of λ_i weighted on the posterior distribution of k. The estimated values are consistent with the parameters used to generate the dataset (Table 3.1)

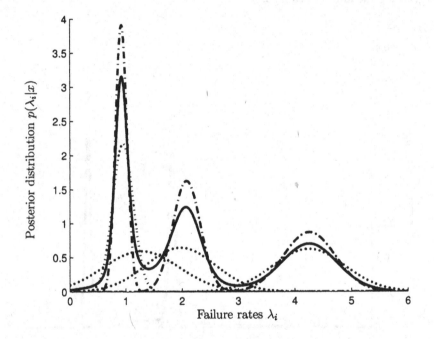

Table 3.1 True and most probable parameter values for the non-homogeneous Poisson model

Parameter	True	Most probable
s_1	84	85.7
s_2	120	121.9
λ_1	1	0.93
λ_2	2	2.05
λ_3	4	4.25

3.4.2 Application to the estimation of the parameters of a deterioration process due to fatigue

Let us now consider a component deteriorating due to fatigue. In these conditions, one of the most popular models used to describe the evolution of crack growth as a function of the stress intensity is the Paris-Erdogan law (see e.g. [31-33] and References therein). The simplest version is given by the equation:

$$\frac{dz(t)}{dt} = az^b(t) \tag{3.23}$$

where $z(t)$ is (proportional to) the crack growth in time (i.e. a suitable measure of the component degradation) and the constants a, b depend on the applied stress range as well as on the thermo-mechanical properties of the probed material. The interval dt is the time span between successive observations of the growing crack.

To account for random effects due to temperature variations, exposure to air, water or chemical reactants, the deterministic law (3.23) can be randomized by introducing a multiplicative noise X_t:

$$\frac{dZ_t}{dt} = aZ_t^b X_t \tag{3.24}$$

Note the change in notation to capital letters, to underline that now the crack growth Z_t is a stochastic process.

The initial condition is set as $Z_0 = z_0 > 0$. As for the multiplicative noise, the functional form $X_t = e^{N(\mu,\sigma)}$ has been assumed, where $N(\mu,\sigma)$ is a Gaussian distribution with mean μ and variance σ^2. It is then convenient to rewrite (3.24) as

$$\frac{dZ_t}{dt} = a_0 Z_t^b X_t \tag{3.25}$$

where the constant a_0 makes the equation dimensionless and the contribution of a in (3.24) is carried out to the parameter μ. Furthermore, the measurement error on the values of Z_t is assumed to be negligible compared to the value of Z_t itself.

The objective is to infer the values of the unknown parameters μ and σ on the basis of a set of measured values of the crack growth Z_{i}, $i = 1, 2, ..., N$. The parameter b is assumed to be known with sufficient accuracy, from previous experimental runs. The constant a_0 can be set to 1 by a suitable change of scale. Moreover, $s_0 = 0$ and $s_{k+1} = T$, the final time at which the evolution of Z_t has been observed, and each experimental run is stopped when Z_t exceeds a fixed threshold $\delta > z_0$.

Observing that $\dfrac{dZ_t}{dt} > 0$ and $Z_t > 0$, the stochastic equation can be recast in the following form:

$$\log(\frac{dZ_t}{dt} Z_t^{-b}) = \log(X_t) \tag{3.26}$$

where the distribution of the stochastic variable at the right-hand side is $N(\mu, \sigma)$.

It is not possible to explicitly derive the distribution of Z_t as a function of the (known) distribution of X_t, which is necessary to obtain the likelihood $L(Z_t | \mu, \sigma)$ for the inference. To overcome this difficulty, Z_t is transformed into the ancillary quantity $Y_t(Z_t) = \log(\frac{dZ_t}{dt} Z_t^{-b})$, whose likelihood function is simply

$$L(Y_t(Z_t) | \mu, \sigma) \propto \frac{1}{\sigma^N} e^{-\sum_{i=1}^{N}(Y_i - \mu)^2 /(2\sigma^2)} \tag{3.27}$$

The reversible-jump MCMC algorithm presented in Section 3.4 can then be used to determine the times s_i of the unknown number k of changepoints at which either μ or σ (or both) have stepwise changes and the corresponding values of the two parameters on each interval between changes.

A numerical example on a synthetic dataset composed of 90 recorded observations is displayed in Fig. 3.8 – Fig. 3.12. The dataset has been generated by introducing a single changepoint for μ (from $\mu_1 = \log(1.2)$ to $\mu_2 = \log(1.6)$) and a single changepoint for σ (from $\sigma_1 = 0.9$ to $\sigma_2 = 0.4$) at two different times ($s_1 = 20$ and $s_2 = 60$) during the experimental run. The changepoints and their locations are not immediately apparent by visual inspection of the recorded data (Fig. 3.8). The reversible-jump algorithm is able to distinguish the actual changepoints locations and to estimate the corresponding parameter values on the subintervals. In Fig. 3.9, the posterior probability mass distribution of the number of changepoints is shown: the most probable value is $k = 2$, but also $k = 3$ and $k = 4$ carry a non-negligible probability.

The conditional probability density functions of the changepoint locations s_i are shown in Fig. 3.10: dotted-dashed curves display the probability density functions conditional to $k = 2$; dotted curves the probability density functions given $k = 3$. The location of a possible third changepoint has been assigned a very broad distribution, denoting a strong uncertainty. The distribution of the changepoint locations weighted on the distribution of k is shown in the same Figure as a solid line. Two peaks are clearly visible and their positions are coherent with the changepoints used to generate the dataset (Table 3.2).

Finally, Fig. 3.11 and Fig. 3.12 show the posterior distributions for the parameters σ_i and $a_i = e^{\mu_i}$: dotted-dashed curves display the distributions conditioned on $k = 2$ and dotted curves the distributions conditioned on $k = 3$. All the plotted curves nearly collapse onto two neat peaks, whose locations almost coincide with the posterior distributions weighted on the distribution of k (solid lines). Again, the distributions are coherent with the values of the parameters used to generate the dataset, as shown in Table 3.2, where the true values of s_i, σ_i and a_i are compared to the most probable values (abscissas of the peaks of the corresponding distributions).

Fig. 3.8 Synthetic dataset for the Paris-Erdogan stochastic model. Abscissas represent time in arbitrary units, ordinates the value of Z_t until a fixed threshold $\delta = 6z_0$ is attained. The component undergoes a step change of μ at $t = 20$ and a second step change of σ at $t = 60$. The two changes are not immediately detectable by simple visual inspection of the component degradation evolution in time

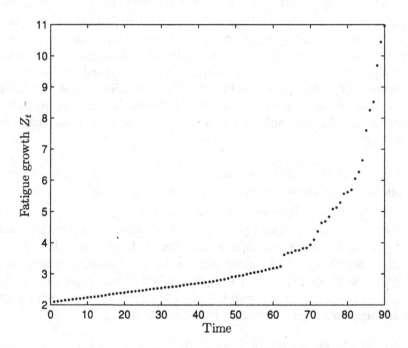

Fig. 3.9 The posterior probability mass distribution of the sampled values
k. The reversible-jump MCMC algorithm has correctly identified that the
most probable number of changepoints is $k = 2$ (the first due to a change of
μ and the second due to a change of σ). The values $k = 3$ and $k = 4$ are
also assigned a non-negligible probability

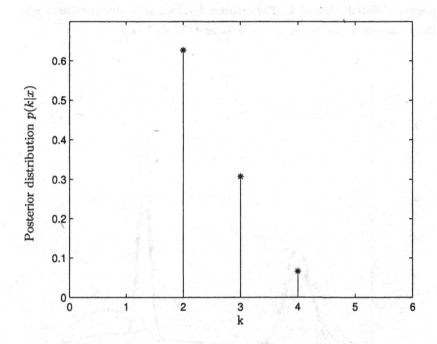

Fig. 3.10 Posterior probability density functions for the locations s_i of the changepoints, conditioned to the number k of changepoints. Dashed-dotted curves represent the distributions conditional to $k = 2$, whereas dotted curves represent the distributions conditional to $k = 3$. The solid line depicts the probability density of the changepoint locations weighted on the posterior distribution of k. The estimated values of s_i are consistent with the parameters used to generate the dataset (Table 3.2)

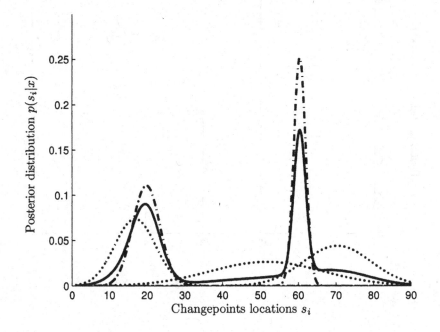

Fig. 3.11 Posterior probability density functions of the parameter σ_i, conditioned to the number k of changepoints. Dashed-dotted curves represent the distributions conditional to $k = 2$, whereas dotted curves represent the distributions conditional to $k = 3$. The solid line depicts the probability density of σ_i weighted on the posterior distribution of k. The estimated values are consistent with the parameters used to generate the dataset (Table 3.2)

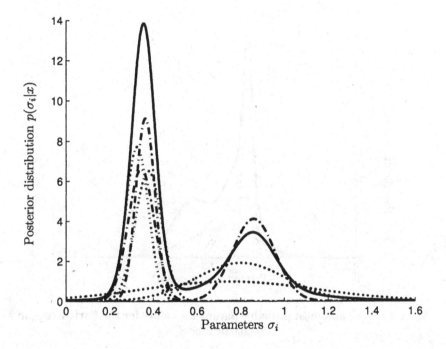

Fig. 3.12 Posterior probability density functions of the parameter $a_i = e^{\mu_i}$ conditioned to the number k of changepoints. Dashed-dotted curves represent the distributions conditional to $k = 2$, whereas dotted curves represent the distributions conditional to $k = 3$. The solid line depicts the probability density of a_i weighted on the posterior distribution of k. The estimated values are consistent with the parameters used to generate the dataset (Table 3.2)

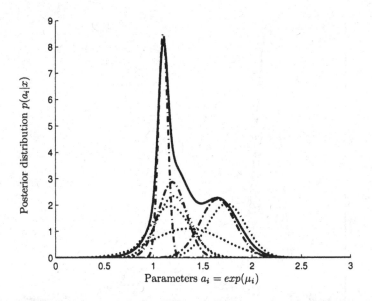

Table 3.2 True and most probable parameter values for the Paris-Erdogan stochastic model

Parameter	True	Most probable
s_1	20	20.5
s_2	60	60.4
σ_1	0.4	0.36
σ_2	0.9	0.86
a_1	1.2	1.11
a_2	1.6	1.65

3.5   Bayesian updating

The Bayesian MCMC inference framework allows incorporating new information as it becomes available. Let us continue considering the component deteriorating according to the Paris-Erdogan fatigue crack growth model of Section 3.4.2 and make the following simplifications:

i)   the dataset has a single changepoint $k = 1$, so that the parameter space has fixed dimensionality and the standard Metropolis-Hastings MCMC algorithm [1-2] can be applied;

ii)   the parameter μ is fixed (and known).

The aim of the study becomes that of identifying the location $T_q = qdt$ of the changepoint and the values σ_1, σ_2 of the parameter σ before and after T_q. The function $\pi(\vartheta \,|\, x)$ is then the posterior distribution of the unknown parameters of the model $\vartheta = [q, \sigma_1, \sigma_2]$, given the information $x = Y_t$. On the basis of the considerations exposed above, the likelihood function $L(x \,|\, \vartheta)$ is taken of the form

$$L(Y_t \,|\, q, \sigma_1, \sigma_2) = \left(\frac{1}{\sigma_1}\right)^q e^{-\sum_{i=1}^{q}(Y_i - \mu)^2/(2\sigma_1^2)} \left(\frac{1}{\sigma_2}\right)^{N-q} e^{-\sum_{i=q+1}^{N}(Y_i - \mu)^2/(2\sigma_2^2)} \tag{3.28}$$

and the prior distribution $p(\vartheta)$ as

$$p(q, \sigma_1, \sigma_2) = U(1, N)N(m_1, s_1)N(m_2, s_2) \tag{3.29}$$

where $U(a, b)$ is the discrete uniform distribution in the interval $[a, b]$ and $N(m, s)$ is the normal distribution with mean m and standard deviation s. The parameter μ in (3.28) is known; the free parameters m_1, m_2, s_1, s_2 are chosen mildly uninformative: a good choice would be to assume s_1, s_2 sufficiently large and to take m_1, m_2 close to the value of

σ_1, which could be known e.g. from previous experimental runs or from the experts' knowledge.

As for the kinetics, the following (symmetric) functional form is taken:

$$K(q',\sigma_1',\sigma_2'|q,\sigma_1,\sigma_2) = U(1,N)N(\sigma_1,d_1)N(\sigma_2,d_2) \qquad (3.30)$$

where the free parameters d_1, d_2 need to be suitably adjusted.

The proposal values $[q',\sigma_1',\sigma_2']$ are accepted (and become the new values in the chain) with probability

$$\alpha = \min\left\{1, \frac{\pi(q',\sigma_1',\sigma_2'|Y_t)}{\pi(q,\sigma_1,\sigma_2|Y_t)} \frac{K(q,\sigma_1,\sigma_2|q',\sigma_1',\sigma_2')}{K(q',\sigma_1',\sigma_2'|q,\sigma_1,\sigma_2)}\right\} \qquad (3.31)$$

Otherwise, they are rejected and the previous values of the chain are retained.

The algorithm is initialized with some random initial values of the chain, $[q,\sigma_1,\sigma_2]^0$. New values $[q',\sigma_1',\sigma_2']$ are then generated by sampling a random number $r \sim U(0,1)$ and checking whether $r \leq \alpha$, in which case the proposal values are accepted and the chain is updated as $[q,\sigma_1,\sigma_2]^1 = [q',\sigma_1',\sigma_2']$. Otherwise, the proposal values are rejected and the chain is updated as $[q,\sigma_1,\sigma_2]^1 = [q,\sigma_1,\sigma_2]^0$. The procedure is repeated until a large chain has been generated: after discarding the burn-in samples, the remaining values approximately obey to the target distribution $\pi(q,\sigma_1,\sigma_2|Y_t)$.

At the end of the first run of the Metropolis-Hastings algorithm, the average value and the standard deviation of the variables $[q,\sigma_1,\sigma_2]$ can be computed. Suppose now that a second dataset becomes available: in the Bayesian framework, a natural way to integrate this new information is to use the previous results as the *a priori* knowledge. More precisely, while in analyzing the first dataset mildly uninformative priors have been adopted, the free parameters m_1, m_2, s_1, s_2 of the prior distribution can now be taken representative of the available prior information.

This can be achieved by setting m_1, m_2 equal to the averages of σ_1, σ_2 estimated above and s_1, s_2 proportional to the standard deviations of σ_1, σ_2. The constant of proportionality is adjusted so to have satisfactory convergence properties: in general, it is not convenient to have very small standard deviations in the prior distribution, since this would lead to a poor exploration of the parameter space. As for the prior distribution for q, a convenient choice is to replace the uninformative $U(1, N)$ with a uniform distribution centered in the average of q estimated at the previous step and with an adjustable width. At the end of the second run of the Metropolis-Hastings algorithm, new estimates for the averages and standard deviations of $[q, \sigma_1, \sigma_2]$ are obtained and these can be used as prior information in a third run of the algorithm, were a new dataset available.

Fig. 3.13 and Fig. 3.14 show the estimates of the average values of the parameters q and σ_1, σ_2, respectively, together with their standard deviations, as a function of the number of Bayesian updates based on the successive datasets examined. The actual (unknown) values of the parameters are also reported as dashed lines: the estimate of q starts with an initial offset due to the first dataset and progressively converges to the real value of the parameter, whereas the estimates of σ_1, σ_2 show very small initial offsets and tend to fluctuate around the real values. The information contained in the successive datasets, thus, is either used to improve (as in the case of q) or confirm (as in the case of σ_1, σ_2) the initial posterior parameter estimates. The efficiency of the proposed algorithm and the speed of convergence strongly depend on the choice of the adjustable parameters, as discussed above.

Fig. 3.13 Posterior estimates (mean and standard deviation) of the parameter $T_q = qdt$ as a function of the number of successive examined datasets. The error bars provide the standard deviations. The dashed line represents the actual (unknown) value of the parameter

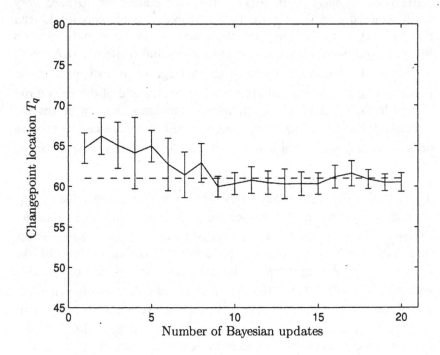

Fig. 3.14 Posterior estimates (mean and standard deviation) of the parameters σ_1, σ_2 **as a function of the number of successive examined datasets. The error bars provide the standard deviations. The dashed lines represent the actual (unknown) values of the parameters**

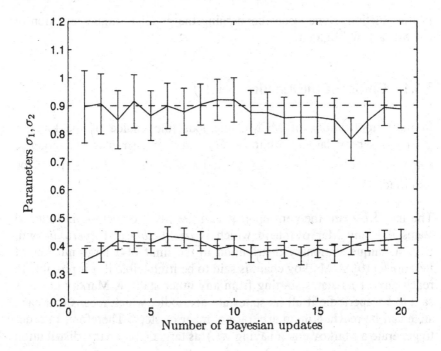

3.6   Practical issues in implementing MCMC algorithms

In this Section, some issues concerning the practical implementation of MCMC are highlighted.

3.6.1   Choice of the kinetics $K\left(.\,|\,.\right)$

In most applications of MCMC methods, the choice of the kinetics $K(.|.)$ has been largely arbitrary. However, to guarantee convergence some weak conditions on the functional form of $K(.|.)$ must be specified.

The most relevant theorem upon which the MCMC method is founded states that any Markov chain which is *irreducible* and *aperiodic* will have a unique stationary distribution in the limit of a large number of iterations [10]. A Markov chain is said to be irreducible if it is possible to reach any of its states starting from any other state. A Markov chain is said to be aperiodic if all its states are aperiodic, which means that each state can be reached in an arbitrary number of steps. Therefore, in order to generate a Markov chain having $\pi(.)$ as target (stationary) distribution one only needs to find a kinetics $K(.|.)$ which satisfies these two properties and the condition $\pi K = \pi$ [10].

Furthermore, a Markov chain with stationary distribution $\pi(.)$ is called *reversible* if the kinetics $K(.|.)$ exhibits the so-called detailed balance $\pi(x)K(y|x) = \pi(y)K(x|y)$. In general, reversibility is a desirable property to have, since any kinetics for which such property is satisfied will admit a stationary distribution $\pi(.)$.

In the previous Sections, the Gibbs sampler and the Metropolis-Hastings algorithm have been considered separately. In practice, however, the kinetics $K(.|.)$ are often constructed by resorting to a combination of different techniques. The choice of the most suitable kinetics (or more generally of their optimal combination) specifically depends on the problem under study.

3.6.2 Burn-in period

One practical problem associated with MCMC is determining the optimal size of the discarded burn-in period, which represents the initial transient phase before convergence of the chain. It is in general very difficult to provide *a priori* an estimate of the rate of convergence to stationarity of the MCMC, as the rates may vary considerably depending on the algorithm chosen and the problems examined. For a detailed review on the empirical numerical methods most frequently used to detect convergence of the chains, the reader is referred to [10].

3.6.3 Number of iterations

Once convergence is attained, it is still necessary to determine how many additional iterations of the algorithm are required to provide a statistically reliable sample. As a general consideration, the number of iterations depends on the examined problem, since the output of the MCMC algorithm is then typically used for Bayesian inference. In this respect, one of the most contentious issues associated with the practical implementation of MCMC algorithms is choosing whether to run one long chain or several shorter chains in parallel [5,10].

The first argument is supported by the fact that after a single long run the chain will be closer to the target distribution, whereas running many short chains implies that several burn-in periods must be discarded. In contrast, the second argument is supported by the consideration that multiple chains run in parallel will explore the sample space much better than a single long chain. This latter approach could also be useful to monitor the mixing properties of the generated sequence by contrasting the different outputs of the chains [10].

3.6.4   Initial conditions

Another relevant issue is the determination of the starting point of the chain, which is typically chosen arbitrarily. By definition, any inference made through MCMC sampling methods will not depend on the starting point of the chain, since all the posterior estimates are drawn on the basis of the values of the chain after equilibrium has been attained (i.e., after the burn-in period has been discarded). However, the choice of the starting point affects the convergence properties of the chain. Several *ad hoc* methods have been proposed for choosing the initial values of the chains; the reader is referred to [10] and References therein for a thorough discussion of the issue.

3.6.5   Other algorithms

In this chapter, three MCMC methods have been illustrated, namely the Metropolis-Hastings, the Gibbs sampler and the reversible-jump algorithms. Although these algorithms are the most widely used in practice, several other schemes have been proposed in order to address specific problems with better overall performances. A detailed review of alternative algorithms may be found in [14] and [10]. Two of the most powerful methods are the continuous time MCMC and the simulated annealing, which will be briefly mentioned here.

In the continuous time MCMC, the Metropolis-Hastings algorithm is adapted by taking into account "search strategies" suggested by the theory of Statistical Mechanics. Jumps defined by the kinetics of the chains are thought to occur at random times which are distributed exponentially. The form of the kinetics is suggested by an analogy with energy gradients and the chain is updated in a way which resembles the diffusion of a particle in an energy well, where the shape of the well is defined by the nature of the examined problem. The updating equation for the kinetics has the structure of a Langevin stochastic transport equation: for these reasons, the evolution of the state of the chain is also called "Langevin diffusion" [14].

The simulated annealing technique provides a way to accelerate the rate of convergence of the MCMC methods by introducing a fictitious "temperature", which is adjusted during the evolution of the chain. The idea behind this scheme comes from Statistical Mechanics, where a diffusing particle explores more frequently the phase space regions with lower energy, whereas the regions at high energy impose a potential barrier to the random walker. In this respect, the temperature of the algorithm may be adjusted in order to achieve a more thorough exploration of the viable space for the states of the Markov chain. Hopefully, the chain will eventually converge to the state characterized by the minimal energy (corresponding to the target equilibrium distribution), avoiding at the same time to get stuck at a local minimum, thanks to a proper choice of the temperature parameter [14].

3.7 References

[1] Metropolis, N., Rosenbluth, A. W., Rosenbluth, M. N., Teller, A. H., Teller, E., *Equations of State Calculation by Fast Computing Machines*, J. Chem. Phys. 21 1087-1091 (1953).

[2] Hastings, W. K., *Monte Carlo Sampling Methods using Markov Chains and their Applications*, Biometrika 57, 97-109 (1970).

[3] Gelfand, A. E., Smith, A.F.M., *Sampling Based Approaches to Calculating Marginal Densities*, J. Amer. Stat. Assoc. 85, 398-409 (1990).

[4] Geman, S., Geman, D., *Stochastic Relaxation, Gibbs Distributions and the Bayesian Restoration of Images*, IEEE Trans. on pattern anal. and machine intell. 6, 721-741 (1984).

[5] Gilks, W. R., Richardson, S., Spiegelhalter, D. J. (Eds.), *Markov Chain Monte Carlo in Practice* (Chapman and Hall, London, UK, 1996).

[6] Smith, A. F. M., Roberts, G. O., *Bayesian Computation via the Gibbs Sampler and Related Markov Chain Monte Carlo Methods (with discussion)*, J. Roy. Stat. Soc. Ser. B 55, 3-24 (1993).

[7] Suparman, S., Doisy, M., Tourneret, J.-Y., *Changepoint Detection using Reversible Jump MCMC Methods*, Proceedings of ICASSP 02 2, 1569-1572 (2002).

[8] Tierney, L., *Markov Chains for Exploring Posterior Distributions (with discussion)*, Ann. Stat. 22, 1701-1762 (1994).

[9] Roberts, G. O., Rosenthal, J. S., *Markov Chain Monte Carlo: Some Practical Implications of Theoretical Results (with discussion)*, Canadian J. Stat. 26, 5-31 (1998).

[10] Roberts, G. O., Rosenthal, J. S., *General State Space Markov Chains and MCMC Algorithms*, Probability Surveys 1, 20-71 (2004).

[11] Brooks, S. P., *Markov Chain Monte Carlo Method and Its Application*, The Statistician 47, 69-100 (1998).

[12] Casella, G., Robert, C. P., *Monte Carlo Statistical Methods* (Springer, 2005).

[13] Kalos, M. H., Whitlock, P. A., *Monte Carlo Methods* (Wiley, New York, 1986).

[14] Geyer, C. J., *Practical MCMC*, Stat. Science 7, 473-483 (1992).

[15] Neal, R. M., *Probabilistic Inference Using Markov Chain Monte Carlo Methods,* Technical Report CRG-TR-93-1 (Dept. of Computer Science, Univ. of Toronto, 1993).

[16] Besaf, J., Green, P., Higdon, D., Mengersen, K., *Bayesian Computation and Stochastic Systems.* Stat. Science 10, 3-66 (1995).

[17] Raftery, A. E., Ackman, V. E., *Bayesian Analysis of a Poisson Process with a Changepoint,* Biometrika 73, 85-89 (1986).

[18] Green, P. J., *Reversible jump Markov chain Monte Carlo Computation and Bayesian Model Determination,* Biometrika 82, 711-732 (1995).

[19] Yang, T. Y., Kuo, L., *Bayesian Binary Segmentation Procedure for a Poisson Process with Multiple Changepoints,* J. Comp. and Graph. Stat. 10, 772-785 (2001).

[20] Ritov, Y., Raz, A., Bergman, H., *Detection of the Onset of Neuronal Activity by Allowing for Heterogeneity in the Changepoints,* J. Neuroscience Methods 122, 25-42 (2002).

[21] Tourneret, J.-Y., Coulon, M., Doisy, M., *Least Square Estimation of Multiple Changepoints Contaminated by Multiplicative Noise using MCMC,* IEEE workshop on Higher-Order Statistics, 148-152, Cesarea, Israel (1999).

[22] Ruggeri, F., Sivaganesan, S., *On Modeling Change Points in Non-Homogeneous Poisson Processes,* Stat. Inf. for Stoch. Proc. 8, 311–329 (2005).

[23] Fearnhead, P., *Exact and Efficient Bayesian Inference for Multiple Changepoint Problems,* Stat. Comp. 16, 203-213 (2006).

[24] C.-Yu Huang and C.-Ti Lin, *Reliability Prediction and Assessment of Fielded Software Based on Multiple Change-Point Models,* Proceedings of PRDC 05, 379-386 (2005).

[25] Bar-Lev, S., Lavi, I., Reiser, B., *Bayesian Inference for the Power-Law Process,* Annals Inst. Math. Science 44, 623-639 (1992).

[26] Ascher, H. E., Feingold, H., *Repairable Systems Reliability* (M. Dekker, New York, 1984).

[27] Rigdon, S. E., Basu, A. P., *Statistical Methods for the Reliability of Repairable Systems* (Wiley, New York, 2000).

[28] Crow, L. H., in Reliability and Biometry, 379-410, (F. Proschan and D. J. Serfling Eds., SIAM, Philadelphia, 1974).

[29] Crow, L. H., *Confidence Interval Procedures for the Weibull Process with Applications to the Reliability Growth,* Technometrics 24, 67-72 (1982).

[30] Erkanli, A., Mazzucchi, T., Soyer, R., *Bayesian Computation for a Class of Reliability Growth Models,* Technometrics 40, 14 (1998).

[31] Sen, A., *Estimation of Current Reliability in a Duane-Based Reliability Growth Model,* Technometrics 40, 334-344 (1998).

[32] Chiquet, J., Eid, M., Limnios, N., *Modelling and Estimating the Reliability of Stochastic Dynamical Systems with Markovian Switching,* Proceedings of ESREL 2006 - Safety and Reliability for managing Risks (2006).

[33] Paris, P. C., Erdogan, F., *A Critical Analysis of Crack Propagation Laws,* J. of Basic Engineering 85, 528-534 (1963).

[34] Heredia-Zavoni, E., Montes-Iturrizaga,, R., *A Bayesian Model for the Probability Distribution of Fatigue Damage in Tubular Joints,* J. Offshore Mech. and Artic Eng. 126, 243 (2004).

4. BASICS OF GENETIC ALGORITHMS WITH APPLICATION TO SYSTEM RELIABILITY AND AVAILABILITY OPTIMIZATION

4.1 Introduction

System reliability and availability optimization is classically based on quantifying the effects that design choices and testing and maintenance activities have on RAM (Reliability, Maintainability, Availability) attributes [1]. A quantitative model is used to assess how the design and maintenance choices affect the system RAM attributes and the involved costs (C). Thus, the design and maintenance optimization problem is framed as a multiple criteria decision making (MCDM) problem where RAM&C attributes act as the conflicting decision criteria with respect to which optimization is sought and the relevant design and maintenance parameters (e.g. redundancy configuration, component failure rates, maintenance periodicities, testing frequencies) act as decision variables or control factors x.

For potentially hazardous and risky industries, such as the chemical and nuclear ones, decision-making for system design and maintenance optimization must account also for risk attributes, which integrate the effects of design and maintenance choices on the system as a whole by including both the likelihood of hazardous events and their expected consequences, e.g. damages to environment and public health, degradation of plant performance, loss of economic income, etc. Correspondingly, the decision-making problem focuses on risk to environmental and public safety, to plant or system performance, to economic assets [2]. For these industries, the system design and maintenance decision-making entails the simultaneous consideration of RAM&Safety (RAMS) criteria [1]. For example, optimization of testing and maintenance activities of safety-related systems aim at increasing their RAM attributes which, in turn, yields an improved plant safety level. This, however, is obtained at the expense of an increased amount of resources invested (e.g. costs, task forces, etc.). Therefore, the multiple criteria decision-making task aims at finding the appropriate choices of reliability design, testing and maintenance procedures that optimally balance the conflicting RAMS and Costs (RAMS&C) attributes. To this aim, the different vector choices of the decision variables $\underline{x}$ are evaluated with respect to numerical objectives regarding reliability, availability, maintainability, risk/safety and cost attributes, e.g. :

- $R(\underline{x})$ = System Reliability;

- $A(\underline{x})$ = System Availability ($U(\underline{x})$= system unavailability = $1 - A(x)$);

- $M(\underline{x})$ = System Maintainability;

- $S(\underline{x})$ = System Safety, normally quantified in terms of the system risk, measure Risk($\underline{x}$) (e.g. as assessed from a Probabilistic Risk Analysis);

- $C(\underline{x})$ = Cost required to implement the vector choice $\underline{x}$.

For different practical and research applications, a reduced decision-making process may suffice based on a subset of the RAMS&C criteria.

In this general view, the vector of the decision variables $\underline{x}$ encodes the parameters related to the inherent equipment reliability (e.g. per demand failure probability, failure rate, etc.) and to the system logic configuration (e.g. number of redundant trains, etc.), which define the system reliability allocation [3-5], and those relevant to testing and maintenance activities (test intervals, maintenance periodicities, renewal periods, maintenance effectiveness, mean repair times, allowed downtimes, etc...), which govern the system availability and maintainability characteristics [6-17].

This chapter discusses the use of Genetic Algorithms within the area of RAMS&C optimization. The theory behind the operation of genetic algorithms is presented. The steps of the algorithm are sketched to some details for both the traditional breeding procedure as well as for more sophisticated breeding procedures. The necessity of affine transforming the fitness function, object of the optimization, is discussed in detail, together with the transformation itself. Finally, two examples of application are illustrated.

4.2 Genetic Algorithms at a glance

Search or optimization algorithms inspired on the biological laws of genetics are called *evolutionary computing algorithms* [18]. The main features of these algorithms are that the search is conducted *i)* using a population of multiple solution points or candidates, *ii)* using operations inspired by the evolution of species, such as *breeding* and *genetic mutation, iii)* based on probabilistic operations, *iv)* using only information on the objective or search function and not on its derivatives. Typical paradigms belonging to the class of evolutionary computing are *genetic algorithms (GAs), evolution strategies (ESs), evolutionary programming (EP)* and *genetic programming (GP).* The focus of this work is on GAs.

As a first definition, it may be said that genetic algorithms are numerical search tools aiming at finding the global maximum (or minimum) of a given real *objective function* of one or more real variables, possibly subject to various linear or non linear constraints [19]. Genetic algorithms have proven to be very powerful search and optimization tools especially when only little about the underlying structure in the model is known. They employ operations similar to those of natural genetics to guide their path through the search space. Essentially, they embed a survival of the fittest optimization strategy within a structured, yet randomized, information exchange [20].

Since the GAs operations are inspired by the rules of natural selection, the corresponding jargon contains many terms borrowed from biology, suitably redefined to fit the algorithmic context. Thus, it is conventional to say that a GA operates on a set of (artificial) *chromosomes*, which are strings of numbers, most often binary. Each chromosome is partitioned in (artificial) *genes*, i.e., each bit-string is partitioned in as many substrings of assigned lengths as the arguments of the objective function to be optimized. Thus, a generic chromosome represents an encoded trial solution to the optimization problem. The binary genes constitute the so called *genotype* of the chromosome; upon decoding, the resulting substrings of real numbers are called *control factors* or *decision variables* and constitute the *phenotype* of the chromosome. The value of the objective function in correspondence of the values of the control factors of a chromosome provides a measure of its 'performance' (*fitness*, in the GA jargon) with respect to the optimization problem at hand.

Fig. 4.1 shows the constituents of a chromosome made up of three genes and the relation between the genotype and the external environment, i.e. the phenotype, constituted by three control factors, x_1, x_2, x_3, one for each gene. The passage from the genotype to the phenotype and viceversa is ruled by the phenotyping parameters of all genes, which perform the coding/decoding actions. Each individual is characterized by a fitness, defined as the value of the objective function f calculated in correspondence of the control factors pertaining to that individual. Thus a population is a collection of points in the solution space, i.e. in the space of f.

Fig. 4.1 Components of an individual (a chromosome) and its fitness

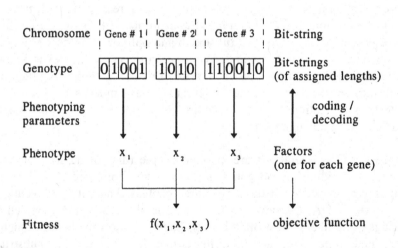

The GA search for the optimal (highest fitness) solution is carried out by generating a succession of *populations* of chromosomes, the individuals of each population being the *children* of those of the previous population and the *parents* of those of the successive population. The initial population is generated by randomly sampling the bits of all the strings. At each step (*generation*) of the search, a new population is constructed by manipulating the strings of the old population in a way devised so as to increase the mean fitness. By analogy with the natural selection of species, the string manipulation consists in selecting and mating pairs of chromosomes of the current population in order to groom chromosomes of the next population. This is done by applying the four fundamental operations of reproduction, crossover, replacement and mutation, all based on random sampling [20]. It has to be noted that in the literature, the two operations of crossover and replacement are often unified and the resulting operation is called crossover, so that the fundamental operations are then three. These operations will be illustrated in some details below. The *evolution* of the chromosomes population is continued until a pre-defined termination criterion is reached.

An important feature of a population, which greatly affects the success of the search, is its genetic diversity, a property which is related to both the size of the population and, as later discussed, to the procedures employed for its manipulation. If the population is too small, the scarcity of genetic diversity may result in a population dominated by almost equal chromosomes and then, after decoding the genes and evaluating the objective function, in the quick convergence of the latter towards an optimum which may well be a local one. At the other extreme, in too large populations, the overabundance of genetic diversity can lead to clustering of individuals around different local optima: then the mating of individuals belonging to different clusters can produce children (newborn strings) lacking the good genetic part of either of the parents. In addition, the manipulation of large populations may be excessively expensive in terms of computer time.

In most computer codes the population size is kept fixed at a value set by the user so as to suit the requirements of the model at hand. The individuals are left unordered, but an index is sorted according to their fitnesses. During the search, the fitnesses of the newborn individuals are computed and the fitness index is continuously updated.

Regarding the general properties of GAs, it is acknowledged that they take a more global view of the search space than many other optimization methods and that their main advantages are:

i) fast convergence to near global optimum
ii) superior global searching capability in complicated search spaces

iii) applicability even when gradient information is not readily achievable.

The first two advantages are related to the population-based searching property. Indeed, while the gradient method determines the next searching point using the gradient information at the current searching point, the GA determines the next set of multiple search points using the evaluation of the objective function at the current multiple searching points. When only gradient information is used, the next searching point is strongly influenced by the local information of the current searching point so that the search may remain trapped in a local minimum. On the contrary, the GA determines the next multiple searching points using the fitness values of the current searching points which are spread throughout the searching space, and it can also resort to the additional mutation to escape from local minima.

4.3 The standard Genetic Algorithm

As said in Section 4.1, the starting point of a GA search is the uniformly-random sampling of the bits of the N (usually an even number) chromosomes constituting the initial population. In general, this procedure corresponds to uniformly sampling each control factor within its range. However, the admissible hypervolume of the control factors values may be only a small portion of that resulting from the cartesian product of the ranges of the single variables, so that conditional sampling may be applied to reduce the search space by accounting for the possible constraints among the values that the control factors can take [21]. In this case, during the creation of the initial population and the successive phase of *chromosome replacement* below described, a chromosome is accepted only if suitable conditioning criteria are satisfied.

Then, the evolution of the chromosomes population from the generic n-th generation to the successive $(n+1)$-th is driven by the so called *breeding algorithm* which develops in four basic steps.

The first step is termed *reproduction* and consists in the construction of a temporary new population. This is obtained by resorting to the following Standard Roulette Selection procedure: *i)* the cumulative sum of the fitnesses of the individuals in the old population is computed and normalized to sum to unity; *ii)* the individuals of the new population are randomly sampled, one at a time with replacement, from this cumulative sum which then plays the role of a cumulative distribution function (cdf) of a discrete random variable (the fitness-based rank of an individual in the old population). By so doing, on the average in the temporary new population individuals are present in proportion to their relative fitness in the old population. Since individuals with large fitness values have more chance to be sampled than those with low fitness values, the mean fitness of the temporary new population is larger than that of the old one.

The second step of the breeding procedure is the *crossover* which is performed as indicated in Fig. 4.2: *i)* $N/2$ pairs of individuals are uniformly sampled, without replacement, from the temporary new population to serve as parents; *ii)* the genes of the chromosomes forming a parents pair are divided into two corresponding portions by inserting at random a separator in the same position of both genes (one-site crossover); *iii)* finally, the head portions of the genes are swapped. By so doing, two children chromosomes are produced, which bear a combination of the genetic features of their parents. A variation of this procedure consists in performing the crossover with an assigned probability p_c (generally rather high, say $p_c \geq 0.6$): a random number R is uniformly sampled in $(0,1]$ and the crossover is performed only if $R < p_c$. Viceversa, if $R \geq p_c$, the two children are copies of the parents.

The third step of the breeding procedure concerns the *replacement* in the new population of two among the four individuals involved in a breeding operation, i.e. the two parents and their two children. The simplest recipe, again inspired by natural selection, simply consists in substituting the parents by their children (children live, parents die). Hence, in this case, each individual breeds only once.

The fourth and last step of the breeding procedure is the random *mutation* of some bits in the population, i.e. the swapping of some bits from their actual values to the opposite one ($0 \rightarrow 1$ and viceversa). The mutation is performed on the basis of an assigned probability of swapping a single bit (generally quite small, say 10^{-3}). The product of this probability by the total number of bits in the population gives the mean number μ of mutations. The bits to be actually mutated are located by uniform random sampling of their positions within the entire bit population. If $\mu < 1$ a single bit is mutated with probability μ.

**Fig. 4.2 Crossover in a population with chromosomes
constituted by three genes**

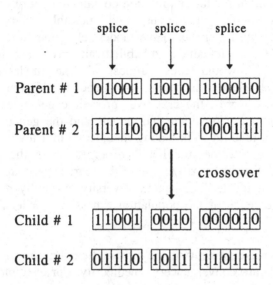

The evolving sequence of successive population generations is usually stopped according to one of the following criteria of search termination:

i.   when the mean fitness of the individuals in the population increases above an assigned convergence value
ii. when the median fitness of the individuals in the population increases above an assigned convergence value
iii. when the fitness of the best individual in the population increases above an assigned convergence value. This criterion guarantees that at least one individual is 'good enough'
iv. when the fitness of the weakest individual in the population drops below an assigned convergence value. This criterion guarantees that the whole population is 'good enough'
v,   when the assigned number of population generations is reached.

4.4 Affine transforming the chromosome fitness

In general, the initial, random population contains a majority of second-rate, i.e. low-fitness, individuals and possibly few mediocre chromosomes which, by chance, have moderately good fitnesses. Then, the standard selection procedure, probabilistically favouring the highest-fitness individuals, would cause almost all the moderately good individuals to be selected as parents and generate children of similar fitnesses. In other words, the crossover procedure generates mediocre individuals substituting mediocre individuals and the genetic selection may be seen as a random walk among mediocres. In a few generations, almost all the second-rate individuals disappear from the population which tends to gather in a small region of the search space around one of the mediocre individuals. The genetic diversity is rapidly reduced and premature convergence of the population fitness onto a local optimum may occur [20].

An appropriate pre-treatment of the fitness function can alleviate this inefficient premature convergence to mediocrity. For example, consider the following affine transform of the fitnesses:

$$f'(x) = a f(x) + b \qquad (4.1)$$

with $a > 0$. Let f_m, f_{avg}, f_M and f'_m, f'_{avg}, f'_M be the minimum, average and maximum fitnesses before and after the transform, respectively. In case of the standard roulette selection procedure, the probability of sampling the k-th individual is $p_k = f_k / \sum_{i=1}^{N} f_i$. If the affine transform is performed, the same individual is sampled with probability

$$p'_k = \frac{f'_k}{\sum_{i=1}^{N} f'_i} = \frac{a f_k + b}{a \sum_{i=1}^{N} f_i + N b} = \frac{p_k + \dfrac{b}{a \sum f_i}}{1 + \dfrac{N b}{a \sum f_i}} = \frac{p_k + \dfrac{1}{N} \dfrac{b}{a f_{avg}}}{1 + \dfrac{b}{a f_{avg}}} \qquad (4.2)$$

Clearly, if b were set to zero, the transform would have no effect on the selection probability. Instead, for $b > 0$, $p_k' > p_k$ if $p_k < 1/N$, that is provided $f_k < f_{avg}$. Therefore, a positive value of b favours the selection of low-fitness individuals. Viceversa, a negative value of b favours the individuals with large fitness values.

Two conditions need to be set to determine the coefficients a and b of the affine transform. One generally adopted condition stems from the following consideration. When going from a generation to the successive one, we must perform N samplings, with replacement, of individuals of the old population to create the N individuals of the new population. In these samplings, the probability of sampling k times the average individual obeys a binomial distribution with probability of success in a single trial $p_{avg} = f_{avg} / \sum f_i$. The expected value of k is then $N \cdot p_{avg} = 1$: on the average, the average individual is sampled once as parent. To preserve this relation after the fitness transform, the original and transformed fitnesses should have equal average, viz.,

$$f_{avg}' = f_{avg} \qquad (4.3)$$

Note that from this condition it follows that the sum of the fitnesses is invariant with respect to the transform, viz.,

$$\sum_i f_i = \sum_i f_i' \qquad (4.4)$$

Therefore, when the parents are selected in proportion to their fitnesses we may compare the probabilities of selecting the k-th individual with reference to the original or to the transformed fitness just by comparing the numerical values of f_k and f_k'.

A second condition follows from the consideration that the evolution of the chromosome population may be divided into two phases, as qualitatively illustrated above:

i.   at the beginning, low-fitness individuals are scattered throughout the search space with few moderately good ones situated by chance in more favourable regions of that space. In this situation, the average population fitness is very low. In the course of the successive generations, the sampling of individuals in proportion to their fitnesses readily drives the population to accumulate around one of the moderately good individuals, with the whole population soon becoming composed of similarly mediocre individuals. A local maximum, most likely mediocre, is thereby converged to. To increase the genetic diversity in this initial phase, one can bias the sampling probabilities so as to favour the breeding of the new individuals by the low-fitness ones which, as above said, are dispersed throughout the search space. This may be achieved by limiting the contribution of the best individual to the population of the next generation. Denoting by C the average number of offsprings bred by the best individual, its value should not be much larger than unity: for typical small populations ranging in the interval from 50 to 100 individuals, a C-value in the range $(1.2, 2)$ is often taken [20]. This second condition yields the following expression for the transform of the maximum fitness:

$$f_M' = C f_{avg} \qquad\qquad (4.5)$$

The a,b values thereby resulting are

$$a = (C - 1)\frac{f_{avg}}{f_M - f_{avg}} \quad ; \qquad\qquad (4.6)$$

The selection of the low-fitness individuals is then favoured provided that b is positive, i.e., for $C < f_M / f_{avg}$. In this case, the few moderately good individuals are scaled down whereas the low-fitness ones are scaled up (Fig. 4.3);

Fig. 4.3 Fitness affine transform

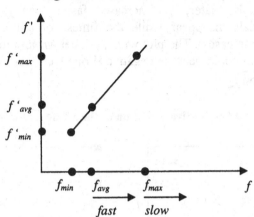

ii.   with the progression of the generations, the fitness of the best individual improves but, in the meantime, the average fitness improves faster getting closer and closer to the best one. If the affine transform is such that a negative value for the minimum transformed fitness, f'_m is attained, which is unacceptable, the second condition is substituted by:

$$f'_m = 0 \qquad\qquad (4.7)$$

and the expressions for the a,b values become:

$$a = \frac{f_{avg}}{f_{avg} - f_m}$$

$$b = -\frac{f_{avg}}{f_{avg} - f_m} f_m \qquad\qquad (4.8)$$

with b becoming negative so that the affine transform turns to favouring the selection of the high-fitness individuals.

Fig. 4.4 qualitatively reports four plots which represent a possible evolution of the affine transform along the generations. In all cases the straight lines are drawn through the two points (f_{avg}, f'_{avg}) and (f_M, f'_M). Note that at the start f_{avg} is close to f_m

(the majority of the population is composed by second-rate individuals); later, f_{avg} increases faster than f_M (second-rate individuals disappear while the fitness of the best individual slowly increases). The plot with the label *wrong* refers to the case in which would become negative if one were not to switch to the condition $f_m' = 0$.

Fig. 4.4 Qualitative evolution of the affine transform

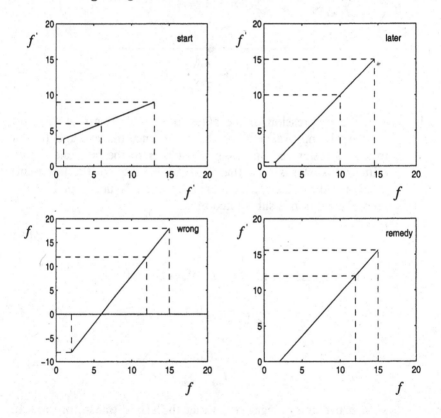

To show the effectiveness of the affine transform, Fig. 4.5 shows a benchmark function to be maximized:

$$z(x,y) = f(x,y) \quad for \quad f(x,y) > 0$$
$$= 0 \quad otherwise \tag{4.9}$$

where

$$f(x, y) = \sqrt{x} \sin(2\pi x) \sqrt{y} \sin(2\pi y) \qquad (4.10)$$

The function is shown to have 50 peaks, with none of them strongly dominating as indicated by the fact that the global maximum (in (5,5) and of value 5) is only 10% higher than the second best one. Therefore, any search algorithm might be likely trapped in a local maximum. A GA maximization technique was applied to this problem, with a population of 100 individuals. The evolution of the first 16 generations are shown in Fig. 4.6, in terms of the histogram of individuals in the population corresponding to the different values of fitness. The upper 16 plots refer to the case with no affine transform of the fitnesses; the lower 16 plots refer to the case of performing the affine transform of the fitnesses. It appears that a maximum is reached earlier in the former case – already at the 11th generation – but the maximum arrived at is only a local maximum. Viceversa, by performing the affine transform, a more thorough exploration of the search space is favoured, thus more efficiently maintaining genetic diversity at the expense of a slowing down of the convergence process but in favour to the attainment of the global maximum, at the 16th generation.

Fig. 4.5 Plot of $z(x,y)$

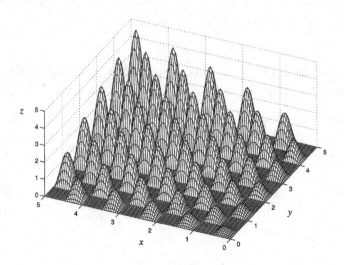

Fig. 4.6 Histograms of the population fitnesses during the first 16 generations. Upper plots: without affine transform; lower plots: with affine transform

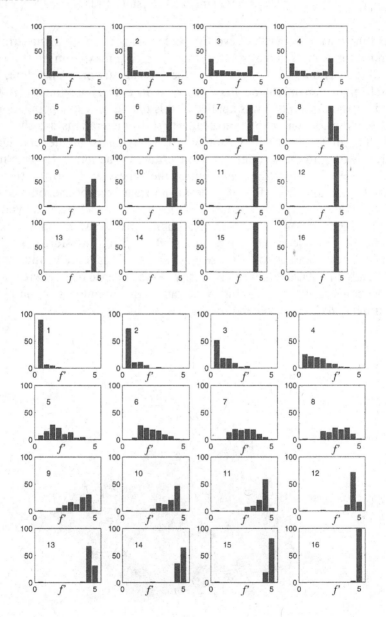

4.5 More sophisticated breeding procedures

Many alternative ways exist to perform the above breeding procedure. In particular, for what concerns the reproduction step, some alternative procedures may be:

i. Hybrid Roulette Selection: the main disadvantage of the Standard Roulette Selection procedure follows from the fact that the new individuals are actually sampled from a multinomial distribution, so that their fitnesses are fairly dispersed around the mean and the convergence towards the best solution can be delayed or even lost. To contrast this, an Hybrid Roulette Selection can be introduced. As in the Standard Roulette Selection case, the procedure starts by normalizing the fitnesses to their sum and by sampling one chromosome from the thereby generated cumulative distribution function. The normalized fitness f_n of the sampled chromosome is then multiplied by the population size N and the integer part of the product yields the number of clones, identical to the sampled chromosome, which are deterministically introduced in the new population. The remainder of the above product is treated as the probability of adding a further identical individual to the new population. This way of proceeding favours the permanence of good individuals, i.e. those with relatively high fitness, along the population generations. However the genetic diversity may worsen.

ii. Random Selection and Mating: the two parents are uniformly, i.e. regardless of the fitnesses of the individuals, randomly selected with replacement, among the entire population. With respect to both the Standard and Hybrid Roulette Methods, this procedure is on the average more disruptive of the parenting genetic codes: in other words, the decodified control factors of the two parent chromosomes may constitute points far apart from each other in the control factor space and, correspondingly, the fitnesses of the newborn children can be largely far apart in the solution space.

iii. Fit-Fit Selection and Mating: the population, rank-ordered on the basis of the individual fitnesses, is scanned and each individual is parent-paired with the next fittest one. On the average, this procedure is highly conservative of the genetic information and a (generally local) maximum of the objective function is soon attained since weak individuals are rapidly eliminated;

iv. Fit-Weak Selection and Mating: as in the preceding case, the rank-ordered population is scanned but this time each individual is parent-paired with the one symmetrically positioned with respect to the mid of the rank-ordered fitness list. On the average, this procedure is highly disruptive of the genetic codes, but it helps in improving the genetic diversity. It is seldom adopted.

In all the described reproduction procedures, after having selected the two parents and before proceeding to the selection of another pair, the two parents are crossed and the two individuals resulting from the adopted replacement procedure are immediately replaced in the population. Most importantly, the population fitness ranking is immediately updated, before sampling the successive pair of parents which is done, then, on a dynamically varying population.

For what concerns the crossover of two parent chromosomes, an obvious generalization of the one-site crossing described above (Fig. 4.2), is the multi-site crossing, consisting in the interposition of more than one separator in the substrings representative of the homologous genes of the parents, followed by the swapping of the pieces of the involved substrings. The simplest case is the two-site crossing: two separators are randomly positioned in the homologous substrings and the bits between the two separator points are interchanged. The multi-site crossing is not commonly adopted so that the simple, one-site, crossover remains by far the most popular technique.

Finally, alternative procedures of replacement are:

i. Fittest individuals: out of the four individuals (two parents and two chromosomes) involved in the crossover procedure, the fittest two replace the parents. This procedure should not be used when weak individuals are largely unfavoured as potential parents in the preceding parent selection step or otherwise they have a significant chance to survive forever in the population.

ii. Weakest individuals: the children replace the two weakest individuals in the entire population, parents included, provided that the children fitness is larger. This technique shortens the permanence of weak individuals in the successive populations and it is particularly efficient in large populations.

iii. Random replacement: the children replace two individuals randomly chosen in the entire population, parents included. By this technique, weak individuals have the same chance as the fit ones of being included in the new population: therefore the method is particularly efficient in small populations since it can give rise to a deep search in the space of the control factors.

4.6 Efficiency of breeding procedures

4.6.1 The figures of merit

The carachteristics of the selection and replacement procedures influence the evolution of the successive generations during the genetic search for the optimum. The efficiency of these procedures can be evaluated with respect to the maximization of analytical functions whose maximum is known in position and value. In particular, the efficiency can be evaluated with respect to:

- the relative error of the maximum value,

$$\varepsilon = \frac{f^* - f}{f^*} \qquad (4.11)$$

where f is the value of the maximum (*fitness*) corresponding to the solution found by the genetic algorithm and f^* is the analytical value of the maximum.

- a measure of the genetic diversity of the chromosomes population or of the archive of best individuals recorded during the search. In this respect, two alternative measures are the standard deviation, σ_{fit}, of the *fitness* of the chromosomes in the population, or in the archive, and the square root of the sum of the variances σ_i^2 of their control factors, $i = 1, 2, ...$

$$d = \sqrt{\sum_i \sigma_i^2} \qquad (4.12)$$

In general, three distinct situations may occur during the genetic evolution:

   i.   convergence of both population and archive;
   ii.   convergence only of the archive;

iii.   convergence only of the population.

The first situation occurs if the search procedure is efficient and one observes the average population and archive fitnesses increasing and correspondingly their standard deviations decreasing, with the successive generations. In the limit of zero variance, the chromosomes in the population and in the archive all converge, hopefully towards the optimal solution.

The second situation that may occur during the search is that the chromosomes in the population do not converge at all so that the population differs greatly from one generation to another. On the other hand, the archive is updated at each generation with the best individuals so that its average fitness is increasing with the successive generations and correspondingly the fitness dispersion decreases while the chromosomes stored in the archive converge, hopefully, to the optimal solution.

Finally, the third, and worse, situation occurs when the population does converge towards an optimum, which is however only local. In this case, the archive is initially filled by individuals dispersed more or less randomly within the search space and then it is not further updated because the population in the successive generations provides only mediocre individuals. In this situation, while the population converges to a local optimum, the archive stabilizes to a dispersed configuration.

Fig. 4.7 shows the three different evolution processes above described with reference to a population of size 500 and an archive containing 400 best individuals. The corresponding evolutions of the standard deviations of the population and archive fitnesses are also shown. The case refers to the maximization of a 5-dimensional Rastrigin function with global maximum in the origin and whose analytical definition will be given later. The search space is the hypercube of side coordinates $[a_j,b_j] = [-10,10], j=1,\ldots,5$.

The first evolution process, in which both population and archive converge, is the most common one and occurs in 11 out of the 4*4 combinations of selection and replacement procedures tested (the Hybrid selection procedure has not been considered here). As explained above, this situation is the one desired as the result of an efficient search.

The replacement procedure *Children→Parents* has been found to lead to the second type of situation, independently of the associated selection procedure, in which the population remains dispersed while the archive converges, although weakly.

The third situation, the most critical one because the genetic algorithm fails to converge to an optimum, occurs when using the combination *Fit-Fit selection, Random replacement*. In this case, in the vicinity of a local optimum the mating of locally optimal parents generates children which most likely belong to the neighbourhood of the same optimum as the parents. Further, with the *Random replacement* the probability of one of the two parents remaining in the population is 10/12, as indicated in Table 4.1 which reports the possible parents pairs (the 10 pairs which give the survival of at least one parent are indicated in bold).

Fig. 4.7 Population and archive evolutions

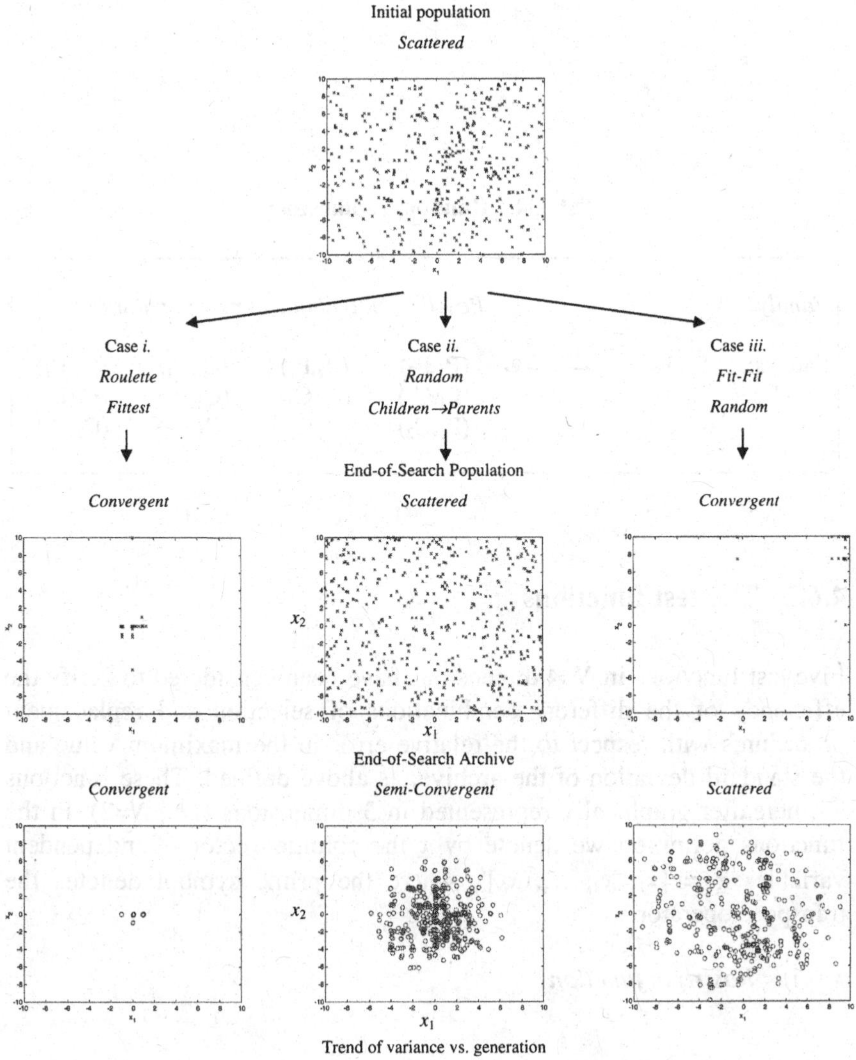

Initial population

Scattered

Case *i.*   Case *ii.*   Case *iii.*

Roulette   *Random*   *Fit-Fit*

Fittest   *Children→Parents*   *Random*

End-of-Search Population

Convergent   *Scattered*   *Convergent*

End-of-Search Archive

Convergent   *Semi-Convergent*   *Scattered*

Trend of variance vs. generation

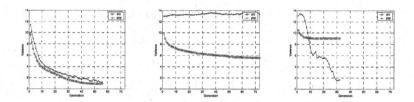

Table 4.1 Random replacement

Family	Possible individuals in new population			
Parents: P_1, P_2	(P_1, P_2)	(P_2, P_1)	(C_1, P_1)	(C_2, P_1)
Children: C_1, C_2	(P_1, C_1)	(P_2, C_1)	(C_1, P_2)	(C_2, P_2)
	(P_1, C_2)	(P_2, C_2)	(C_1, C_2)	(C_2, C_1)

4.6.2   The test functions

Five test functions in $N=4$ dimensions have been considered to verify the efficiency of the different combinations of selection and replacement procedures with respect to the relative error in the maximum value and the standard deviation of the archive, as above defined. These functions are hereafter graphically represented in 3 dimensions (i.e., $N=2$). In the functions definition we denote by $\underline{x}$ the column vector of independent variables $\underline{x} = [x_1, x_2, ..., x_N]'$, where the prime symbol denotes the transpose operator.

   i)   Rastrigin function

$$f(x) = 110N - \sum_{i=1}^{G} \left(x_i^2 - 10\cos\left(2px_i\right) \right) \quad \text{with } x_i \in [-10,10], \ i=1,...,N=4$$

Fig. 4.8 Rastrigin function

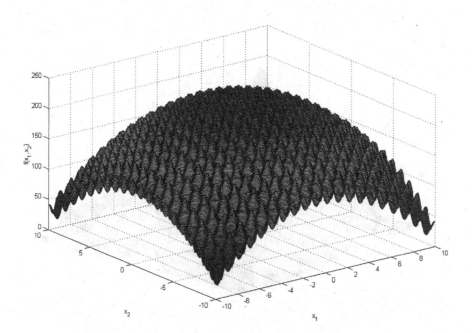

Rastrigin function [22] is symmetric with respect to all its variables and presents multiple local peaks, with a global one in the origin of the coordinates equal to $f(\underline{0})=120N$. The local maxima of first order have all coordinates equal to zero except one whose value is $x_i=1$, $i=1,...,N=4$: this renders very difficult the search for the global maximum.

ii) Michalewicz function

$$f(\underline{x}) = S_i\left(\sin(x_i)\sin^{2m}\left(i\frac{x_i^2}{\pi}\right)\right) \text{ with } m = 10, x_i \in [0,\pi], i = 1,...,N = 4$$

Fig. 4.9 Michalewicz function

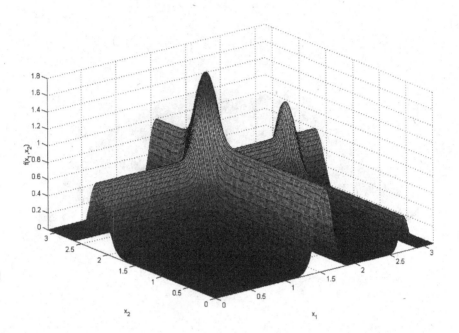

Michalewicz function [23] is a multimodal function with $N!$ local optima. The parameter m defines the steepness of the valleys and peaks. As m increases, the difficulty of the search increases. In the limit, for very large values of m, the maxima tend to become Dirac-δs and the points outside the peaks neighbourhoods give very limited information on the position of the global optimum. For $N=4$ and $m=10$, the position of the maximum is $\underline{x}^* \approx [2.203\ 1.571\ 1.285\ 1.923]'$ and its value is $f(\underline{x}^*) \approx 3.699$.

iii) Product of modulated sines

$$f'(\underline{x}) = \prod_{i=1}^{N} \sqrt{x_i} \sin\left(2\pi x_i\right)$$

$$f(\underline{x}) = \begin{cases} f'(\underline{x}) & \text{if} \quad f'(\underline{x}) > 0 \\ 0 & \text{otherwise} \end{cases} \qquad x_i \in [0,5], \, i = 1,\dots,N = 4$$

Fig. 4.10 Product of modulated sines

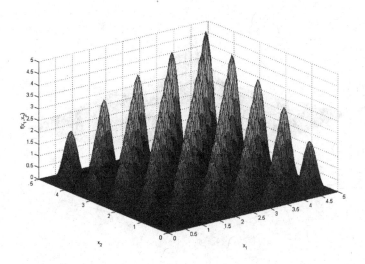

For N even, this function is symmetric with respect to its variables and the only global maximum $f\left(\underline{x}^{*}\right) = \sqrt{\left(x_i^{*}\right)^{N}}$ is obtained when all the coordinates are equal to their maximum value $x_i^{*} = \dfrac{2N+1}{2} + \dfrac{1}{N}$ (for N=4, x_i^{*}=4.75), i=1,...,4. The secondary maxima of higher order are modulated by the square root and equidistant. This function is similar to Rastrigin's but the presence of zones where it remains constant render the search more difficult because the few points which belong to such zones are equivalent in terms of fitness, independently of their actual distance from the global maximum.

iv) Langermann function

$$s_k = S_i \left(x_i - a_{ik} \right)^2$$

$$f(\underline{x}) = 200N - S_k c_k \exp\left(-\frac{s_k}{\pi} \right) \cos(ps_k)$$

$$with\ k = 1,...,M = 2, x_i \in [-10,10],\ i = 1,...,N = 4$$

Fig. 4.11 Langermann function

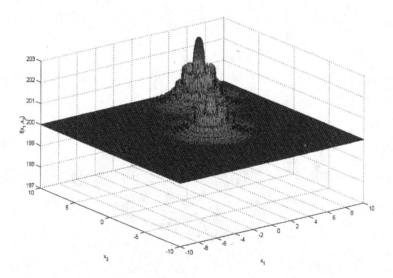

This function presents M maxima at $x_i = a_{ik}$, $i=1,...,N$, $K=1,...,M$ whose importance is determined by the constants c_k. The maxima are modulated by the product of a cosine, which gives them an oscillatory character, and a Gaussian, which damps these effects as one moves away from the center of the coordinate space. In the genetic algorithm searches which follow, we considered $M=2$ maxima, one of which located in the origin of the coordinate axes and the other at $\underline{a_2}=[7\ 5\ 3\ 2]'$ and $c_1 = -2$, $c_2 = -3$. Similarly to Rastrigin function, the difficulty in searching for the global maximum of this function stems from the presence of large areas in which f is constant and whose points give no information on the location of the maximum.

v) Schwefel function

$$f\left(\underline{x}\right) \ = \ 100\,N + S_i x_i \sin\left(\sqrt{|100x_i|}\right) \quad x_i \in [0,10],\ i=1,...,N=4$$

Fig. 4.12 Schwefel function

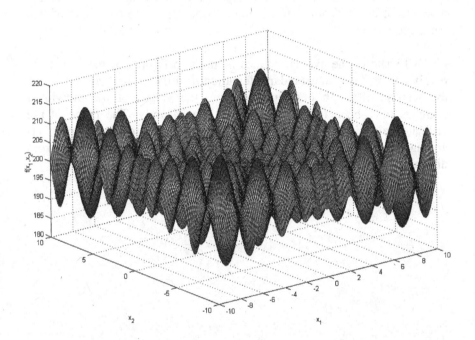

Schwefel function [24] presents various local maxima and a single global one, $f = 435.63$, in correspondence of a vector of coordinates all equal to $x_i = -\left(\dfrac{19}{20}\pi\right)^2$, $i=1,...,4$. The local maximum closest to the global one is actually well separated, at $x_i = \left(\dfrac{17}{20}\pi\right)^2$, $i=1,...,4$, and has a value of approximately 428.52. The function is therefore somewhat tricky and the search for the global optimum quite difficult.

4.6.3 Results

The five test functions presented above have been used to evaluate the efficiency of different combinations of selection and replacement procedures, with reference to the relative error in the maximum value and the standard deviation of the archive. For each combination, various genetic algorithm searches have been repeated for different values of the initial random seed, to test the robustness of the results, and different numbers of chromosomes in the population, to analyze the effect of this parameter.

Fig. 4.13 shows the average values of the two figures of merit, i.e. the relative error ε and the standard deviation d in the archive at the end of the search. Although not reported here, the values of the individual searches have been found to not differ significantly. The range of the ordinate values is not the same in all graphs, to ease the comparisons among the different procedures on the same function rather than across different functions.

Fig. 4.13 Relative error and standard deviation

Relative error ε Dispersion d

Rastrigin function

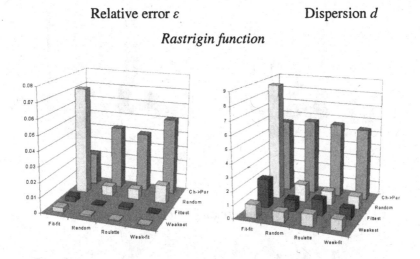

Michalewicz function

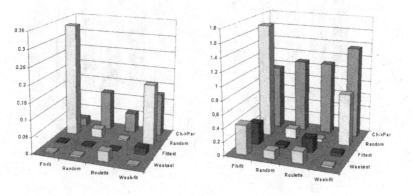

Product of modulated sines

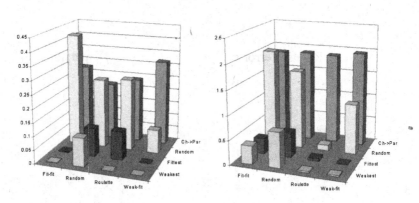

Langermann function

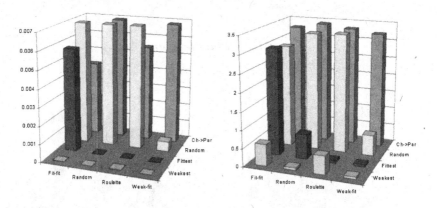

Schwefel function

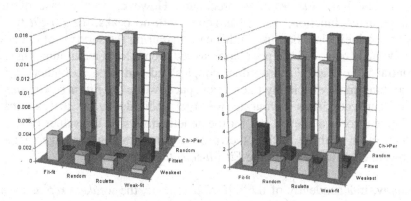

From the graphs it appears evident that the *replacement* procedure has a more significant effect on the search than the *selection* one.

When the *children→parents* procedure is used, the relative error is large since the children replace the parents in the population independently of the respective fitnesses. For the search to be efficient with respect to this figure of merit, it is necessary to combine this 'fitness-free' replacement procedure with a 'fitness-guided' parents *selection* procedure. In particular, the *Fit-Fit*, which is the most strongly 'fitness-guided' procedure, leads to the smallest relative error, although its value is generally larger than those which are obtained using the other *replacement* procedures. As for the other figure of merit, *d*, it is rather large independently of the associated *selection* procedure (see also case ii. of the previous Fig. 4.7).

The *random replacement* procedure is certainly one which favours genetic diversity (case iii., Fig. 4.7) at the expense of the goodness of the optimum found. The results of the searches performed with this procedure are dependent on the function object of the optimization. Clearly, the combination of *random replacement* and *Fit-Fit selection*, leads to large values of both the relative error and the dispersion figures of merit (case iii. of Fig. 4.7).

When the replacement procedure promotes the fittest individuals, the population efficiently moves towards the maximum, thus achieving small relative errors. The smallest relative error is found in combination with a *Fit-Fit* or *Weak-Fit selection* procedure. However, in the case of the Langermann function, which has zones with no peaks, the *Fit-Fit* turns out to be inefficient because in this case favouring the survival of the best individuals may lead the population to settle on a local maximum. On the contrary, the *Weak-Fit* procedure couples bad and good individuals thus searching more efficiently the search space, while the *fittest replacement* procedure continues favouring the best individuals, so that a small relative error is achieved. As for the genetic diversity, the values of *d* are comparable to those of the other procedures, with slightly larger values when in combination with the *Fit-Fit selection* procedure.

Finally, independently of the *selection* scheme, the *weakest replacement* procedure is very efficient with respect to the relative error, since it allows to rapidly get rid of the bad individuals in the population. This however goes at the expense of the genetic diversity which is rapidly lost: all the population is driven to convergence on an optimum which may very well be only local.

Tables 4.2 and 4.3 and Figures 4.14 and 4.15 report values of the relative error and dispersion averaged over the five test functions considered. These values confirm the previous findings, i.e.:

1. the *replacement* procedure is more relevant than the *selection* one, with respect to both the relative error and the dispersion;
2. some combinations are particularly inefficient (especially the *Fit -Fit + Random*);
3. to obtain a small relative error, it is necessary to perform a *replacement* based on the fitness of the chromosomes involved (*Fittest* and *Weakest*);
4. to increase genetic diversity, the *replacement* must be "natural" (*children→parents*) or *random*.

Table 4.2 Average relative error

Fit-Fit	Random	Roulette	Weak-Fit	Error
0.25%	2.28%	0.65%	0.09%	*Weakest*
16.99%	6.29%	6.08%	5.88%	*Random*
6.61%	7.91%	7.15%	10.39%	*Ch→Par*
0.30%	2.20%	2.17%	0.47%	*Fittest*

Fig. 4.14 Relative error

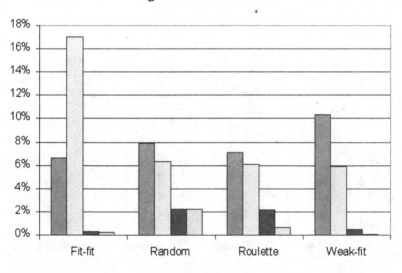

Table 4.3 Average dispersion

Fit-Fit	Random	Roulette	Weak-Fit	Dispers.
1.64	0.55	0.66	0.77	Weakest
5.46	3.45	3.02	2.46	Random
4.80	5.03	5.02	4.99	Ch→Par
1.85	0.65	0.33	0.34	Fittest

Fig. 4.15 Average dispersion

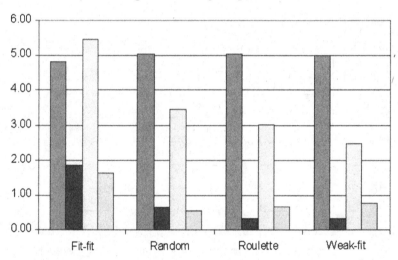

■ Ch->Par □ Random ■ Fittest □ Weakest

4.7 Inducement of species and niches

GAs can be further extended to benefit from the inclination of natural species to forming groups and mixing among groups, a behaviour which has proven of utmost advantage for the survival and the development of the species.

In the present context, a *species* is defined as a class of individuals sharing common features, a *niche* is a set of functions performed by the individuals of a species and the *environment* is the collection of the external conditions, including the interactions with the other species. Examples of species, niche and environment are the humans, the elephant hunters and the ivory market, respectively. In nature, we observe a large number of species which live - possibly competing - simultaneously. They develop or decline according to their degrees of adaptability to the environmental changes and survive until the environment is propitious to them. In other words, each species has found a niche and survives until the niche is favourable.

In the following, we shall briefly describe only three important techniques which derive from these considerations: the Isolation by Distance, the Spatial Mating and the Sharing [20]. In all cases, the selection of the parents of the next population is performed within a limited number of individuals and therefore with a remarkable saving of computer time.

4.7.1 Isolation by distance

The population of chromosomes is divided in sub-populations, i.e. in groups of individuals with scarce mutual interactions. Adopting the jargon of demography, it is said that the individuals of each sub-population live in isolated *islands* and that the various islands evolve separately, with different convergence rates, towards possibly different solutions. Such a rigid picture of totally isolated islands may be softened by the introduction of *emigrants*: in general, the individuals living in an island mate with individuals of the same group, but there is also a small number of wanderlust individuals, the emigrants, which travel to other islands looking for the ideal mate.

The algorithmic implementation of the Isolation by Distance procedure is straightforward: the sub-populations in the various islands reproduce separately following the rules described in the preceding Sections for the whole population. Additionally, an emigration matrix E is assigned, whose generic element E_{ij} gives the emigration probability of an individual from island i to island j. Let us consider a particular island: at the beginning of each reproduction step, the numbers of immigrants from the other islands is sampled from the probability matrix E and the individuals selected as below specified are added to the natives. In the meantime an equal number of native elements (generally the weakest ones) are eliminated in order to keep constant the population size. The selection of the individuals emigrating from an island may be done in several different ways, e.g.:

i.   at random from the current sub-population. This choice results in a great mixing of genes which prevents an excessive similarity among the individuals of an island;

ii. by sending copies of the fittest individual to the other islands. With respect to the previous procedure, this choice is more guided and therefore the genetic diversity among the different sub-populations is lower.

4.7.2   Spatial mating

The population is again divided in sub-populations called *demes* but these now overlap almost totally. A deme is one of the administrative divisions of the ancient Attica and of modern Greece. A deme is created around each individual, by specifying a user-assigned number of neighbours so that there is an equal number of individuals and demes. In the reproduction phase of the new population creation, one of the two parents is still selected from the entire population following one of the procedures previously illustrated, but its mating partner is now selected within the corresponding deme only. Several definitions of the deme topology are possible:

i.   the individuals are disposed on a planar square grid and the deme of
     an individual is constituted by its immediate neighbours
     (8,24,48,...);
ii. the individuals are disposed on a monodimensional wheel and the
     deme of an individual is constituted by an equal number of
     individuals symmetrically disposed around it (2,4,6,...);
iii. the deme is the entire population, again disposed on a
     monodimensional wheel: the second parent is selected by means of
     a random walk game with a given interval probability, generally an
     exponential distribution with a mean value of few steps: in practice
     the deme has a boundary of a few mean free paths.

4.7.3 Sharing

In nature, when the environment in which a species lives starts to become
crowded, a survival possibility is that the individuals agree to share the
available resources. To do this, each individual which takes possession of
some good must share that good among the other individuals.

This behaviour can be implemented within a GA search procedure by
modifying the fitnesses upon which the selection of the parents and their
replacement are performed in such a way that the natural fitness of each
individual is decreased in relation to its similarity with the remaining
individuals in the population. To do this the following steps must be
performed:

i.   definition of a *distance* between pairs of individuals. Possible
     distances Δ_{ij} between chromosomes i and j are:

     -   distance between phenotypes:

$$\text{Euclidean distance} \quad \Delta_{ij} = \sum_{k=1}^{G} \left(x_{ik} - x_{jk} \right)^2 \qquad (4.13)$$

$$\text{Chebyshev distance} \quad \Delta_{ij} = \sum_{k=1}^{G} | x_{ik} - x_{jk} | \qquad (4.14)$$

where G is the number of genes in a chromosome and x_{rk} is the control factor of the k-th gene in the r-th chromosome;

- distance between genotypes:

$$\text{Hamming distance} \quad \Delta_{ij} = \sum_{k=1}^{B} H_{ij}^{k} \qquad (4.15)$$

where B is the number of bits in a chromosome and

$$H_{ij}^{k} = 1 \quad \text{if chromosomes } i,j \text{ have equal } k\text{-th bit}$$
$$= 0 \quad \text{otherwise}$$

ii. introduction of a *sharing function* $s(\Delta_{ij})$ between chromosomes i and j, defined in such a way that $s(0)=1$ and $s(\Delta_{ij})=0$ for $\Delta_{ij} \geq \beta$, where β is a user's defined value: in words, a chromosome shares 1 with himself and 0 with chromosomes farther away than β. Often, a linear $s(\Delta)$ is adopted.

iii. Computation of the modified fitness $f_i^{'}$ of the i-th chromosomes:

$$f_i^{'} = \frac{f_i}{\sum_{j=1}^{N} s(\Delta_{ij})} \qquad (4.16)$$

The main advantage of the inducement of niches by the various methods is the partition of the whole population in subpopulations which evolve towards the various maxima of the objective function.

4.8 Multi-objective optimization

The multi-objective optimization problem arises when in correspondence of each point x in the search space, one must consider several objective functions $f_i(x)$, $i = 1,2,...$, possibly conflicting. The comparison of two solutions with respect to several objectives can be achieved by introducing the concepts of *Pareto optimality* and *dominance* [25,26] which enable solutions to be compared and ranked without imposing any a priori measure as to the relative importance of individual objectives, neither in the form of subjective weights nor arbitrary constraints.

Consider N different objective functions $f_i(x)$, $i = 1,2,...,N$. Solution x *dominates* solution y if x is better on all objectives:

$$f_i(x) > f_i(y) \quad for\ i=1,...,N \tag{4.17}$$

The solutions not dominated by any other are *nondominated* solutions.

Within the genetic approach, in order to treat simultaneously several objective functions it is necessary to generalize the single-fitness procedure used in the single-objective GA by assigning N fitnesses to each x.

Concerning the insertion of an individual (an x value) in the population, constraints often exist which impose restrictions that the candidate individual has to satisfy and whose introduction speeds up the convergence of the algorithm, due to a reduction in the search space. Such constraints can be handled, just as in the case of single-objective GAs, by testing whether, in the course of the population creation and replacement procedures, the candidate solution fulfills the criteria pertaining to all the N fitnesses.

Once a population of individuals (chromosomes) $\{x\}$ has been created, we rank them according to the Pareto dominance criterion by looking at the N-dimensional space of the fitnesses $f_i(x)$, $i=1,2,...,N$, (see Fig. 4.16 for $N=2$). All nondominated individuals in the current population are identified. These solutions are considered the best ones, and assigned the rank #1. Then, they are virtually removed from the population and the next set of nondominated individuals are identified and assigned rank #2. This process continues until every solution in the population has been ranked.

Fig. 4.16 Example of population ranking for a problem of maximization of f_1 and f_2

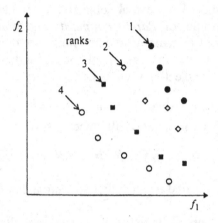

The selection and replacement procedures of the multiobjective genetic algorithms are based on this ranking: every chromosome belonging to the same rank class has to be considered equivalent to any other of the class, i.e. it has the same probability of the others to be selected as a parent and survive the replacement.

During the optimization search, an archive of vectors, each one constituted by a nondominated chromosome and by the corresponding N fitnesses, representing the dynamic Pareto optimality surface is recorded and updated. At the end of each generation, nondominated individuals in the current population are compared, in terms of the fitnesses, with those already stored in the archive and the following archival rules are implemented:

1. If the new individual dominates existing members of the archive, those dominated members are removed and the new one is added;
2. if the new individual is dominated by any member of the archive, it is not stored;

3. if the new individual neither dominates nor is dominated by any member of the archive then:

   a. if the archive is not full, the new individual is stored.
   b. if the archive is full, the new individual replaces the *most similar* one in the archive (an appropriate concept of distance being introduced to measure the similarity between two individuals: this paper adopts a Euclidean distance based on the values of the fitnesses of the chromosomes normalized to the respective mean values in the archive).

The archive of nondominated individuals is also exploited by introducing an elitist parents' selection procedure which should, in principle, be more efficient. Every individual in the archive (or, if the archive's size is too large, a pre-established fraction) is chosen once as a parent in each generation. This should guarantee a better propagation of the genetic code of nondominated solutions, and thus a more efficient evolution of the population towards Pareto optimality.

At the end of the search procedure the result of the optimization is constituted by the archive itself which gives the Pareto optimality region.

4.9 Application of genetic algorithms to RAMS

The design, operation and management of an engineered system or plant requires proper accounting of the constraints coming from safety and reliability requirements as well as from budget and resource considerations.

At the design stage, analyses are to be performed in order to guide the design choices in consideration of the many practical aspects which come into play and which typically generate a conflict between safety requirements and economic needs: this renders the design effort an optimization one, aiming at finding the best compromise solution. In particular, typically the reliability design optimization problem regards a choice among alternative system configurations made up of components which possibly differ for their failure and repair characteristics. The safety vs. economics conflict rises naturally as follows:

Choice of components: choosing the most reliable components certainly allows the design to be on the safe side and guarantees a high system availability but it may be largely non-economic due to excessive component purchase costs; on the other hand, less reliable components provide for lower purchase costs but loose availability and may increase the risk of costly accidents.

Choice of redundancy configuration: choosing highly redundant configurations, with active or standby components, increases the system reliability and availability but also the system purchase costs (and perhaps even the repair costs, if the units composing the redundancy are of low reliability); obviously, for assigned component failure and repair characteristics, low redundancies are economic from the point of view of purchase costs but weaken the system reliability and availability, thus increasing the risk of significant accidents and the system stoppage time. These very simple, but realistic, aspects of plant design immediately call for compromise choices which optimize plant operation in view of its safety and budget constraints.

The above problem is referred to as the *redundancy allocation problem* and it has been shown to be NP-hard [27]. Various optimization approaches have been proposed to solve the problem, such as dynamic programming, integer programming, mixed integer and nonlinear programming and various heuristics. A summary of the earliest applications of these methods to series-parallel systems is presented in [28,29].

In spite of the realistic need of achieving several objectives (e.g. low costs, high revenues, high reliability, low accident risks), some of which in conflict, and at the same time satisfying several requirements (e.g. maximum allowable weight, volume etc.), typically these approaches tackle the problem as a single-objective optimization formulated in the following two ways [25]:

1) weighed aggregation of all the objectives into a single objective function to be optimized;
2) optimization of one of the objectives by itself and imposition on the others of appropriate constraints to be satisfied.

Both these single-objective approaches are inevitably open to criticism, the former one because of the homogenization of different quantities, such as reliability, costs, health consequences, to a common unit of measure and the latter one because of the choice of which objective function to optimize and of the setting of the constraints levels.

Since the mid-nineties, genetic algorithms have been successfully applied for the solution of single-objective redundancy optimization problems under multiple constraints [30-36]. The genetic algorithm considers a population of chromosomes, each one encoding a different alternative reliability design solution. Each design solution proposed by the genetic algorithm during the search is evaluated with respect to the objective function. The evaluation can be done analytically, under simplifying assumptions, or by Monte Carlo simulation, for a more realistic representation of the system reliability behaviour (chapter 2). In the latter case, a Monte Carlo code should be run for each individual of the chromosome population throughout all the generations. For realistic systems, this would lead to excessive computing times. A possible solution to this problem follows from the consideration that in the genetic algorithm approach the best chromosomes appear a large number of times in the successive generations whereas the bad ones are readily eliminated. Then, for each proposed chromosome, one can run a Monte Carlo code with a limited number of trials, e.g. 500, obtaining poorly significant statistical results. An archive of the simulated configurations, and corresponding Monte Carlo estimates, is maintained and updated by discarding the least fit configuration when the archive dimensions are exceeded. Then, whenever a chromosome is re-proposed, the results thereby obtained are accumulated with those stored in the archive as obtained in previous Monte Carlo runs pertaining to the same chromosome. The large number of times a good chromosome is proposed in the successive generations allows accumulating over and over the results of the few-histories runs, thus achieving at the end statistically significant results. At the same time, this way of proceeding avoids wasting time on 'bad' configurations which have small fitness values and are therefore simulated only a small number of times. This approach has been termed 'drop-by-drop' for its similarity to this way of filling a glass of water [13].

Apart from series-parallel systems, network design problems are becoming increasingly critical and complex as telecommunication networks (and others) are expanded and upgraded in response to consumers' needs. Indeed, recently there has been much significant research considering network reliability optimization. The problem of the assignment of the proper reliability level to the links of a network with fixed topology, as a means to optimize the system reliability, is an NP-hard problem [37]. Different methods exist for tackling this problem in various applications related to network systems. In general, the focus is on single objective problems, e.g., maximization of the system reliability. Single-objective genetic algorithms also have been applied to determine the optimal design network configuration by selecting components from multiple alternatives [38-40].

Further, as explained in Section 7, a more informative approach to system design optimization is the multi-objective one which considers all individual objectives separately, aiming at identifying a set of solutions better than others with respect to all objectives, but 'comparatively good' among themselves. Each member of this set is better than or equal to the others of the set with respect to some, but not all, of the objectives. The set thereby identified provides a spectrum of 'good' solutions which the decision maker can subjectively handle according to which objectives he believes to be more or less important. For example, between two solutions a decision maker could prefer the one with highest reliability although obtained at higher costs or vice versa he might privilege low costs, thus giving up some reliability.

Several papers have been published to address redundancy allocation of series-parallel systems under multiple objectives. For example, in [41] a multiple-criteria optimization approach has been applied to maximize system reliability and minimize resources consumption (cost, weight and volume). In [42,43] integer programming and a min-max concept have been used for obtaining Pareto optimal solution. In [9], genetic algorithms have been applied to find a set of Pareto optimal solutions in the reliability design of nuclear safety systems.

For network system design, there has been only limited research considering multiple objectives.

Concerning the effective management of an engineered system or plant, this strongly relies on diagnostics, surveillance and maintenance procedures. An efficient strategy in this respect can ensure both safe operation and economic gain; on the contrary unsatisfactory performances and substantial waste of resources may result from an inefficient management of these activities.

Few applications of single-objective genetic algorithms exist in the area of fault diagnostics, both in classical series-parallel systems [44] and in network systems [45,46].

Significantly more promising is the use of genetic algorithms for the optimization of maintenance and surveillance. Particular focus has been given to the determination of the optimal time intervals for the periodic testing of the system components and to replacement scheduling strategies, in order to maximize availability and minimize costs. Both single-objective and multi-objective approaches have been proposed [1,9,13,15,47-50]. On condition maintenance has also been considered. For example, in [51] a multi-objective search is framed, based on a combination of genetic algorithms and Monte Carlo simulation, for determining the optimal 'on-condition' maintenance strategy in terms of the thresholds of components degradation beyond which maintenance has to performed. The optimal threshold of degradation for each component is searched with respect to the system mean availability over the mission time and the system net profit within the mission time. The genetic algorithm considers a population of chromosomes, each one encoding a threshold degradation value for each component type. For a given chromosome, the Monte Carlo simulation estimates the two objective functions.

Finally, genetic algorithms are also efficiently used for the estimation of the effective parameters of simplified (reduced) models of complex dynamics [52-54]. The calibration of the effective parameters is achieved by best fitting the model responses of the quantities of interest to the actual evolution profiles. This is achieved by constructing appropriate objective functions. The use of fast, reduced models for the study of system dynamic behavior is widely spread in safety design since they provide, in reasonable computing times, a qualitative understanding of the physical phenomena involved. Furthermore, the rapidity of the calculations of these models allows the performance of sensitivity and uncertainty analyses and render more feasible a dynamic approach to reliability analysis [55,56]. Clearly, safety-critical transients still need to be simulated by more detailed models which capture more accurately the phenomena involved.

4.10 Examples

4.10.1 Multi-objective optimization of system design: a simple application [13]

When designing a system such as an industrial plant, one must give proper account to the constraints coming from the safety and reliability requirements as well as from the limitations on budget and resources. In particular, the problem here considered regards the choice among different potentially valid system redundancy configurations made up of components which can differ for their failure and repair characteristics. The choice of a higher redundancy or of a more reliable component certainly increases the safety properties of the plant, but also increases the investment costs. On the contrary, less reliable components lead to a reduction of the purchase cost, but also a reduction of system availability and an increase in the risk of costly accidents. Thus, conflicting aspects in the plant design call for an optimal compromise choice between safety and budget requirements.

To illustrate the features of the multiobjective approach, we consider a simple system modeled by two nodes in series. The optimization problem deals with the choice of the redundancy allocation for each node and of the types of components to be used. We can choose 1-out-of-nG (Good) parallel configurations, with $n=1,...,4$ for the first node and $n=1,...,3$ for the second node. The choice of the types of components to be used in the system is made through the choice of the corresponding failure rates within pre-defined ranges (this is done obviously for the sake of simplicity as in practice the selection of components is done out of a set of few available components characterized by different, discrete values of failure rate). Given the methodological purpose of this first example, for simplicity we assume that the failure rate λ_i^k, expressed in y^{-1}, is the only parameter involved in the description of the performance of each component: all the other quantities characterizing the component are given in relation to its failure rate. Hence, the repair rate of component i in node k is related to its failure rate as follows:

$$\mu_i^k = \alpha\sqrt{\lambda_i^k} \tag{4.18}$$

where α is a pre-defined constant parameter, expressed in $\dfrac{1}{\sqrt{y}}$.

The conjecture behind this relationship is that a more reliable component is technologically more complicated and will therefore require longer restoration times (and, as we shall see below, greater costs). We also assume, for the sake of simplicity, that the number of repairmen is equal to the number of components constituting the system.

Two separate objectives of the optimization problem are considered: the net profit drawn from system operation during the mission time (T_M) and the reliability at mission time.

The profit is made up of the following contributions [57]:

- profit from plant operation P;
- purchase and installation costs C_A;
- repair costs C_R;
- penalties during downtime, due to missed delivery of agreed service C_{NS}.

The net profit objective function G (gain) can then be written as follows:

$$G = P - (C_A + C_R + C_{NS}) \tag{4.19}$$

where

$$P = P_t \cdot \int_0^{T_M} A(t)dt \tag{4.20}$$

is the plant profit in which P_t represents the amount of money per unit time paid by the customer for the plant service and $A(t)$ is the instantaneous plant availability.

$$C_A = \sum_{k=1}^{N_n} \sum_{i=1}^{n_k} C_{A,i}^k \tag{4.21}$$

is the acquisition and installation cost of the N_n nodes, the k-th of them constituted of n_k components; $C_{A,i}^k = \dfrac{\gamma^k}{\sqrt{\lambda_i^k}}$ is the contribution due to component i in node k and γ^k $\left(\dfrac{\$}{\sqrt{y}}\right)$ is a proportionality constant equal for all components of the k-th node.

$$C_R = \sum_{k=1}^{N_n}\sum_{i=1}^{n_k} \overline{C_{R,i}^k} \qquad (4.22)$$

is the mean repair cost of all components of the system, with $\overline{C_{R,i}^k}$ being the mean value of repair cost for component i in node k. Such mean repair cost is assumed to obey the following model:

$$\overline{C_{R,i}^k} = N_{R,i}^k \left(\frac{1}{\mu_i^k}\right) \cdot C_{R,i}^k \qquad (4.23)$$

where:

$C_{R,i}^k = \dfrac{\beta^k}{\mu_i^k}$ (yearly repair cost for component i in node k; β^k is a proportionality constant equal for all components of the k-th node, expressed in $\dfrac{\$}{y^2}$)

$N_{R,i}^k = \dfrac{T_M}{\dfrac{1}{\lambda_i^k}+\dfrac{1}{\mu_i^k}}$ (mean number of failures during the mission time)

and finally

$$C_{NS} = C_U \cdot \int_0^{T_M} [1 - A(t)]dt \qquad (4.24)$$

in which C_U is the economic penalty per unit time, i.e. the amount of money to be paid to the customer because of missed delivery of the agreed service when the plant is unavailable.

Note that in this single model we did not introduce the interest rates. The second objective function considered is the reliability at mission time $R(T_M)$. The evaluation of this objective is performed simply through the resolution of the markovian process governing the stochastic evolution of the system considered.

Concerning the genetic algorithm, one gene encodes the system configuration and one gene is used for the failure rate of each component, which are seven at most, so that the chromosome is made up by eight genes which give a complete description of a potential system solution to the optimization problem.

Results and discussion

Table 4.4 contains the parameters related to the system technical and economical specifications, whereas Table 4.5 contains the rules and the parameters for the GA implemented in order to solve the two-objectives optimization problem with the in-house developed Fortran code MOGA (MultiObjective Genetic Algorithm - http://lasar.cesnef.polimi.it/). The values of the parameters were chosen based on experience and trial-and-error tuning, so as to achieve proper convergence.

The complete search space resulting from the failure rate ranges and the other system parameters is shown in Fig. 4.17, in which the values of the two objective functions for all possible solutions are represented.

Table 4.4 Economic parameters of case study 1

Mission time (T_M)	30 (y)
Ranges of failure rates (λ_i^k) for node $k=1$	$[10^{-3},10^{-1}]y^{-1}$
Ranges of failure rates (λ_i^k) for node $k=2$	$[10^{-2},10^{-1}]y^{-1}$
Parameter linking failure and repair rates	10
Parameters for acquisition and installation cost: – Node 1: γ^1 – Node 2: γ^2	85 \$ $y^{-1/2}$ 80 \$ $y^{-1/2}$
Parameters for repair cost – Node 1: β^1 – Node 2: β^2	500 \$ y^{-2} 520 \$ y^{-2}
Profit per unit time (P_t)	100 \$/y
Penalty due to plant downtimes (C_U)	200 \$/y

Table 4.5 Genetic Algorithm rules and parameters

Number of genes per chromosome	8
Number of bits per gene –	7
Number of chromosomes (population size, N_p)	100
Number of generations (termination criterion)	400
Mutation probability	0.001
Selection technique	Standard Roulette
Number of generations without elitist selection	40
Replacement technique	Weakest
Number of nondominated chromosomes in the archive	400

The sharp changes in the shape of the search space are due to the configuration changes of the system nodes occurring when n in the 1-out-of-nG configuration varies.

Fig. 4.17 Search space for the multiobjective optimization problem

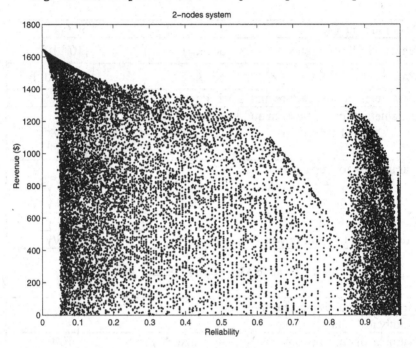

The results of the GA optimization process are shown in Fig. 4.18. It can be seen how the GA efficiently identifies the Pareto optimal solutions, i.e. the nondominated solutions. The discontinuities in the Pareto surface are due to changes in the system configuration which introduce jumps in the achievable values of the system reliability. For example, there are no nondominated solutions with reliability values in the range between 0.45 and 0.85.

Based on this information the decision maker can either impose a minimal reliability level for the plant as an a posteriori constraint, or he can decide to sacrifice part of the system revenues in favor of an increased system reliability. Actually, looking at the high reliability solutions, a natural option in our case would seem that of obtaining higher revenues with a little decrease in reliability. In any case, it is shown that adopting a multiobjective approach provides a wider information to the decision maker without introducing any 'a priori' arbitrariness which, instead, comes into play 'a posteriori', at the decision level.

Fig. 4.18 Multiobjective optimization results: revenue vs. reliability

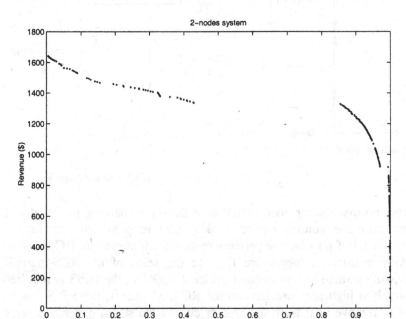

4.10.2 Multi-objective optimization of the inspection policy of a nuclear safety system [8]

Let us consider a standby safety system of a nuclear power plant (NPP) [50,58]. The system under consideration is the high pressure injection system (HPIS) of a pressurized water reactor (PWR). Fig. 4.19 shows a simplified schematics of a specific HPIS design. The system consists of three pumps and seven valves.

Fig. 4.19 The simplified HPIS system

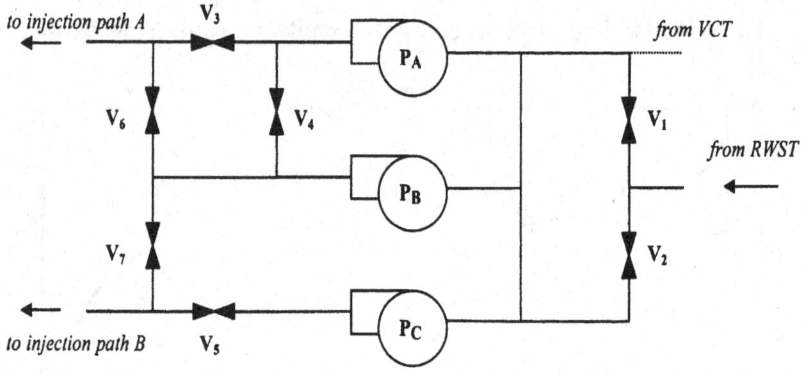

RWST = Radioactive Waste Storage Tank

During normal reactor operation, one of the three charging pumps draws water from the volume control tank (VCT) in order to maintain the normal level of water in the primary reactor cooling system (RCS) and to provide a small high-pressure flow to the seals of the RCS pumps. Following a small loss of coolant accident (LOCA), the HPIS is required to supply a high pressure flow to the RCS. Moreover, the HPIS can be used to remove heat from the reactor core if the steam generators were completely unavailable. Under normal conditions the HPIS function is performed by injection through the valves V_3 and V_5 but, for redundancy, crossover valves V_4, V_6 and V_7 provide alternative flow paths if some failure were to occur in one of the nominal paths.

This stand-by safety system has to be inspected periodically to test its availability. The test interval (TI) specified by the technical specifications (TS) both for the pumps and the valves is 2190 h. However, there are several restrictions on the maintenance procedures described in the TS, depending on reactor operations. For this study the following assumptions are made:

1. At least one of the flow paths must be open at all times.
2. If the component is found failed during surveillance and testing, it is returned to an as-good-as-new condition through corrective maintenance or replacement.

3. If the component is found to be operable during surveillance and testing it is returned to an as-good-as-new condition through restorative maintenance.
4. The process of inspection and testing requires a finite time; while the corrective maintenance (or replacement) requires an additional finite time, the restorative maintenance is supposed to be instantaneous.

Moreover, in this study the system components have been divided in three groups characterized by different test strategies. All the components belonging to a same group undergo testing with the same periodicity. The groups identified through the test period T^i, $i=1,2,3$, are :

$$T^1 \rightarrow V_1, V_2$$
$$T^2 \rightarrow P_A, P_B, P_C, V_3, V_5$$
$$T^3 \rightarrow V_4, V_6, V_7$$

Assuming a mission time of one year, the range of variability of the three TIs is [0,8760] hours. Therefore, any solution to the optimization problem can be encoded using the following array of decision variables: $\underline{x} = \{T^1, T^2, T^3\}$

Problem formulation and objective functions

The goal is to optimize the effectiveness of the TIs of the HPIS with respect to three different criteria:

i)   mean availability
ii) cost
iii) workers' time of exposure to radiation.

The TIs then represent the decision variables of the optimization problem and different choices of their values will lead to different performances with respect to the three above mentioned objectives.

Mean unavailability

To compute the system unavailability, the fault tree for the top event "no flow out of both injection paths A and B" has been developed. The boolean reduction of the corresponding structure function allows us determining the N system minimal cut sets (MCS). Then, the mean system unavailability $\overline{U}$ can be expressed as [50]:

$$\overline{U} \approx \sum_{j=1}^{N} \prod_{i=1}^{n_j} \overline{u}_i^{\,j} \qquad (4.25)$$

where N is the number of MCS, n_j is the number of basic events relevant to the j-th minimal cut set and $\overline{u}_i^{\,j}$ represents the mean unavailability associated with the i-th component belonging to the j-th MCS.

As for the mean unavailability $\overline{u}_i$ of a generic individual component i, several models have been proposed in the literature to account for the different contributions coming from failure on demand, human errors, maintenance etc. In this study, the following model is assumed [50]:

$$\overline{u}_i = \rho_i + \frac{1}{2}\lambda_i\tau_i + \left(\rho_i + \lambda_i\tau_i\right)\frac{d_i}{\tau_i} + \frac{t_i}{\tau_i} + \gamma_0 \qquad (4.26)$$

where:

ρ_i = probability of failure on demand
λ_i = failure rate of i-th component
τ_i = test interval of i-th component
t_i = mean downtime due to testing
d_i = mean downtime due to corrective maintenance
γ_0 = probability of human error

Equation (4.26) is valid for $\rho < 0.1$ and $\lambda\tau < 0.1$ which are reasonable assumptions when considering safety systems.

Obviously, the adopted model of Eqs. (4.25) and (4.26) is a practical but simplified model: a more realistic approach would require the use of Monte Carlo simulation for the evaluation of the system unavailability [13,57].

Cost function

We assume that the cost objective C is made up of two major contributions:

i) $C_{S\&M}$ = costs associated with surveillance and maintenance ($S\&M$);

ii) $C_{accident}$ = costs associated with consequences related to accidents possibly occurring at the plant, therefore,

$$C = C_{S\&M} + C_{accident} \qquad (4.27)$$

For a given component i the $S\&M$ costs are computed on the basis of given yearly inspection ($C_{ht,i}$) and corrective maintenance ($C_{hc,i}$) costs.

For a given mission time, T_M, the number of inspections performed on component i are $\dfrac{T_M}{\tau_i}$; of these, on average a fraction equal to $(\rho_i + \lambda_i \tau_i)$ demands also a corrective maintenance action. Thus, the surveillance and maintenance costs amount to:

$$C_{S\&M} = \sum_{i=1}^{N_C} \left[C_{ht,i} \left(\frac{T_M}{\tau_i} \right) t_i + C_{hc,i} \left(\frac{T_M}{\tau_i} \right) d_i \left(\rho_i + \lambda_i \tau_i \right) \right] \qquad (4.28)$$

As for what concerns the accident costs contribution, $C_{accident}$, this is intended to measure the costs associated to damages of accidents which are not mitigated due to the HPIS failing to intervene. A proper analysis of such costs implies that we account for the probability of the corresponding accident sequences. To this aim, the small LOCA event tree of Fig. 4.20 has been considered [34].

Fig. 4.20 Event tree for the initiating event small LOCA [34]

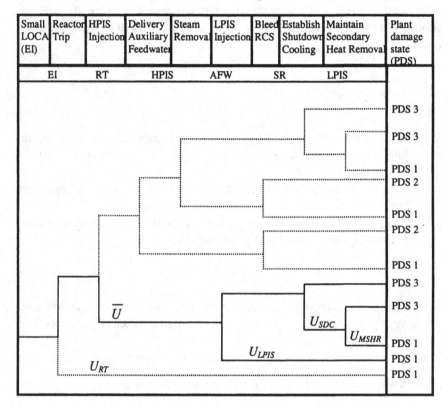

Actually, the HPIS plays an important role in many other accident sequences generating from other initiators such as intermediate LOCA, station blackout, turbine trip etc. In this illustrative example, only the contribution due to small LOCAs is considered, recognizing that by so doing the resulting accident cost contribution related to the HPIS is significantly underestimated. Table 4.6, also taken from [34], reports the characteristics of the plant damages states (*PDSs*) resulting from the various small LOCA accident sequences, and the economic damages of the associated consequences.

Table 4.6 Description of possible plant damage states associated to the initiating event small LOCA [34]

PDS	Plant damage state	Health risk	Investment risk	Total risk
1	Severe core damage or core melt; significant radioisotope release to containment	$5*10^4$ person rem per event ($\sim U(5,50)$ M\$/event) median: 27.5 M\$	$\sim U(1162,3136)$ M\$/event median: 2149 M\$	$\sim U(1167,3186)$ M\$/event median: 2176.5 M\$
2	Small LOCA leading to containment cleanup, valve and vessel repair to containment	$3.8*10^4$ person rem per event ($\sim U(3.8,38)$ M\$/event) median: 20.9 M\$	$\sim U(329,924)$ M\$/event median: 626.5 M\$	$\sim U(332.8,962)$ M\$/event median: 647.4 M\$
3	Possible damage to steam generator; minor containment cleanup and equipment checkout		$\sim U(32,243)$ M\$/event median: 137.5 M\$	$\sim U(32,243)$ M\$/event median: 137.5 M\$
4	Possible primary system water loss; little or no spill into containment; no core or equipment damage		$\sim U(1,6)$ M\$/event median: 3.5 M\$	$\sim U(1,6)$ M\$/event median: 3.5 M\$

The accident sequences considered for the quantification of the accident costs are those which involve the failure of the HPIS (solid lines in Fig. 4.20), so that the possible *PDS* are *PDS*1 and *PDS*3. Thus:

$$\begin{cases} C_{accident} = C_1 + C_3 \\ C_1 = P(EI) \cdot (1 - U_{RT}) \cdot \overline{U} \cdot \{U_{LPIS} + (1 - U_{LPIS}) \cdot U_{SDC} \cdot U_{MSHR}\} \cdot C_{PDS1} \\ C_3 = P(EI) \cdot (1 - U_{RT}) \cdot \overline{U} \cdot (1 - U_{LPIS}) \cdot \{U_{SDC} \cdot (1 - U_{MSHR}) + (1 - U_{SDC})\} \cdot C_{PDS3} \end{cases}$$

$$(4.29)$$

where C_1 and C_3 are the total costs associated with accident sequences leading to damaging states 1 and 3, respectively. These costs depend on the initiating event frequency and on the unavailability values of the safety systems which ought to intervene along the various sequences: these values are taken from the literature [34,59] for all systems except for the SDC and MSHR, which were not available and were arbitrarily assumed of the same order of magnitude of the other safety systems, and for the HPIS for which the unavailability is calculated from Eqs. (4.25) and (4.26) and it depends on the test intervals of the components. Finally, for the values of C_{PDS1} and C_{PDS3}, the accident costs for $PDS1$ and $PDS3$ respectively, the mean values of the uniform ditributions given in Table 4.4 were adopted. Table 4.7 summarizes the input data.

Table 4.7 Frequencies associated to failures of safety systems in the ET and costs associated to PDSs of interest for sequences involving the HPIS failure

Frequency of small LOCA (y^{-1}) [34]	$2.43 \cdot 10^{-5}$
Frequency of Reactor Trip failure (y^{-1}) [60]	$3.6 \cdot 10^{-5}$
Frequency of LPIS failure (y^{-1}) [60]	$9 \cdot 10^{-3}$
Frequency of SDC failure (y^{-1})	$5 \cdot 10^{-3}$
Frequency of MSHR failure (y^{-1})	$5 \cdot 10^{-3}$
Mission time (h)	8760
Cost associated to PDS 1 ($\$ \cdot$event^{-1}) = C_{PDS1}	$2.1765 \cdot 10^{9}$
Cost associated to PDS 3 ($\$ \cdot$event^{-1}) = C_{PDS3}	$1.375 \cdot 10^{8}$

Exposure time

During testing operations, the technicians may be subjected to radiation exposure. With reference to the ICRP recommendation n° 60 [61], based on the well known ALARA (As Low As Reasonably Achievable) and limit-dose principles, the dose received by workers should be minimized. Assuming a constant exposure rate, the minimization of the dose is equivalent to that of the exposure time, so that the third objective function of our optimization problem can be assumed to be:

$$T_{exp} = \sum_{i=1}^{N_c} \left[\left(\frac{T_M}{\tau_i} \right) t_i + \left(\frac{T_M}{\tau_i} \right) d_i \left(\rho_i + \lambda_i \tau_i \right) \right] \qquad (4.30)$$

with the same meaning of the symbols explained in the previous subsections.

Note that, for simplicity, in computing the radiation exposure time with Eq. (4.30) we neglect the fact, often verified in practice, that work management procedures in nuclear power plants are such that exposure times associated with performing surveillance tests or corrective maintenance on a component are larger than its respective mean downtimes.

Expression (4.30) is similar to that of Eq. (4.28) for the surveillance and maintenance costs, $C_{S\&M}$. However, the presence of the accident contribution in the cost objective function is such that exposure time and cost are generally two distinct objectives to be optimized separately.

An analysis of the three objective functions hereby defined shows that they all share some common contributions but present some conflicting ones as well. For example, the cost function has a contribution relating to the unavailability of the HPIS due to economic damages of occurring accidents and a contribution associated to the time of surveillance and maintenance (and thus of exposition) due to the costs of such operations. On the other hand, the surveillance and maintenance time influences also the mean system unavailability, through the downtimes of the inspected components.

Genetic coding

The goal of the work is that of utilizing the multiobjective genetic algorithm optimization procedure to determine the optimal values of inspection intervals, T^i, i=1,2,3 for the three groups of components identified in the HPIS, which maximize separately the three objective functions: average availability ($\overline{A} = 1 - \overline{U}$), reciprocal of the cost (C^{-1}) and reciprocal of the exposure time ($T_{\exp}^{-1}$). The decision variables of the optimization are then the three test intervals T^i, i=1,2,3. Such test intervals are assumed to vary in the range [1, 8760]h so that at least one inspection on each component is carried out in one year. Each of the variables is coded by one 10-bit gene in the chromosome. The data relevant for the multiobjective genetic algorithm procedure contained in Table 4.8 have been selected after appropriate tuning and constitute the input to the MOGA code.

Table 4.8 Genetic Algorithm parameters and rules

Number of chromosomes (population size, N_p)	100
Number of generations (termination criterion)	500
Selection	Standard Roulette
Replacement	Weakest
Mutation probability	0.005
Crossover probability	1
Number of non-dominated chromosomes in the archive	400

Multiobjective optimization of the HPIS

Fig. 4.21 shows the results obtained through the genetic algorithm procedure for maximizing the three objective functions of mean unavailability, reciprocal of costs and reciprocal of exposure time, simultaneously. In the Figure, we report the values of the objective functions in correspondence of all the nondominated solutions (triplets of TIs) contained in the archive at convergence.

It is clear that there exists a linear relationship between cost and exposure time. This is due to the fact that the safety systems failure frequencies and accidental costs are such that the contribution to cost due to accidents is negligible compared to that of surveillance and maintenance, which, in turn, is proportional to the surveillance and maintenance time and, thus, to exposure time. As a consequence of this linear relation, the results obtained would not change, as it was verified, if we were to perform a two-objective optimization in which the two objectives were the mean availability and one between the reciprocals of cost and exposure time. Along these lines, an optimization considering the reciprocal of cost and exposure time as the only objective functions would result in a single-point solution, corresponding to the lowest cost and exposure time.

Fig. 4.21 Multi-objective optimization results

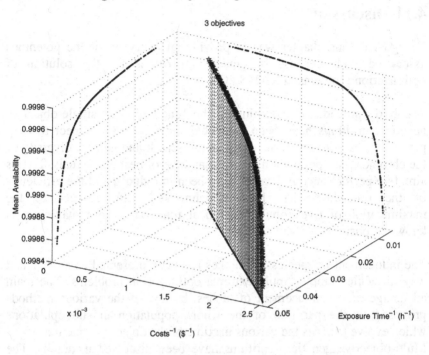

Finally, the test intervals in the Genetic Algorithm's archive give an indication that the HPIS can indeed be made more available, on average, by increasing the frequency of the inspections but, as reasonable, this leads to large inspectors' exposure times and also renders the system more expensive. A thorough analysis of the results in the archive also shows that T^1 is somewhat dominant, as expected since it governs the inspections on the two valves V_1 and V_2 which constitute the most critical MCS of the system.

4.11 Discussion

The goal of this chapter was to provide an overview of the potentials associated with the use of genetic algorithms for the solution of optimization problems in RAMS&C.

The basic procedures underpinning the functioning of single-objective genetic algorithms have been presented critically with respect to their performances in terms of accuracy of the search and genetic diversity of the chromosomes population or of the archive of best individuals. To this aim, two performance indicators have been introduced and the efficiency of the procedures has been benchmarked with respect to the maximization of five popular analytical functions whose maximum is known in position and value.

The inducement of species and niches has been presented as an efficient way to tackle complex multi-extrema optimization problems. The main advantage of the inducement of niches, by any of the various methods proposed, is the partition of the whole population in subpopulations which evolve towards the various maxima of the objective function.
Multi-objective genetic algorithms have been discussed in details. The multi-objective point of view is more informative in that it allows identifying the set of Pareto optimal solutions which can be provided to the decision makers for further analysis and evaluation according to their subjective preference values.

Finally, the genetic algorithms have been discussed with respect to their use in RAMS&C analysis, particularly in the area of redundancy allocation, fault diagnostics, maintenance and surveillance optimization.
Although the discussion is by no means exhaustive and the references cited do not pretend to cover all the work done in the area, the issues analyzed show the great potentials associated with this optimization technique. Particularly, the combination of the genetic algorithm search with the Monte Carlo simulation evaluation of the objective functions seems very promising for the optimization of systems and plants under realistic conditions. However, further work needs to be done in the actual implementation of these methods in practical situations, which may bring additional challenges and open new issues of research and development.

4.12 References

[1] Martorell, S., Villanueva, J.F., Carlos, S., Nebot, Y., Sanchez, A., Pitarch, J.L. and Serradell, V., *RAMS+C Informed Decision-Making with Application to Multi-Objective Optimization of Technical Specifications and Maintenance using Genetic Algorithms*, Reliability Engineering and System Safety, 2005, Vol. (87), pp. 65-75.

[2] Frank, M., *Choosing among Safety Improvement Strategies: a Discussion with Example of Risk Assessment and Multi-Criteria Decision Approaches for NASA*, Reliability Engineering and System Safety, 1995, Vol. (49), pp. 311-324.

[3] Painton, L. and Campbell, J., *Genetic Algorithms in Optimization of System Reliability*, IEEE Transactions on Reliability, 1995, Vol. 44(2), pp. 172-178.

[4] Cantoni, W., Marseguerra, M. and Zio, E., *Genetic Algorithms and Monte Carlo Simulation for Optimal Plant Design*, Reliability Engineering and System Safety, 2000, Vol. 68 (1), pp. 29-38.

[5] Levitin, G. and Lisnianski, A., *Join Redundancy and Maintenance Optimization for Multistage Series-Parallel Systems*. Reliability Engineering and System Safety, 1999, Vol. 64 (1), pp. 33-42.

[6] Vaurio, J.K., *Optimization of Test and Maintenance Intervals based on Risk and Cost*, Reliability Engineering and System Safety, 1995, Vol. 49 (1), pp. 23-36.

[7] Muñoz, A., Martorell, S. and Serradell, V., *Numerical Absolute & Constrained Optimization of Maintenance based on Risk and Cost Criteria using Genetic Algorithms*, Proceedings of Advances in Safety and Reliability, ESREL 1997, Lisbon, Portugal, pp. 1749-1756.

[8] Harunuzzaman, M. and Aldemir, T., *Optimization of Standby Safety System Maintenance Schedules in Nuclear Power Plants*, Nuclear Technology, 1996, Vol. 113 (3), pp. 354-367.

[9] Busacca, P.G., Marseguerra, M. and Zio, E., *Multi-objective Optimization by Genetic Algorithms: Application to Safety Systems*. Reliability Engineering and System Safety, 2001, Vol. 72 (1), pp. 59-74.

[10]   Cepin, M., *Optimization of Safety Equipment Outages Improves Safety*, Reliability Engineering and System Safety, 2002, Vol. 77 (1), pp. 71-80.

[11]   Martorell, S., Carlos, S., Sanchez, A. and Serradell, V., *Simultaneous and Multi-Criteria Optimization of TS Requirements and Maintenance at NPPs*, Annals of Nuclear Energy, 2002, Vol. 29 (2), pp. 147-168.

[12]   Muñoz, A., Martorell, S. and Serradell, V., *Genetic Algorithms in Optimizing Surveillance and Maintenance of Components*, Reliability Engineering and System Safety, 1997, Vol. 57 (2), pp. 107-120.

[13]   Marseguerra, M. and Zio, E., *System Design Optimization by Genetic Algorithms*, Proceedings of the Annual Reliability and Maintainability Symposium, RAMS 2000, January 24-27, Los Angeles, CA, pp.222–227.

[14]   Lapa, C.M.F., Pereira, C.M.N.A. and Melo, P.F.F.E., *Surveillance Test Policy Optimization through Genetic Algorithms using Non-Periodic Intervention Frequencies and considering Seasonal Constraints*, Reliability Engineering and System Safety, 2003, Vol. 81 (1), pp. 103-109.

[15]   Tsai, Y.T., Wang, K.S. and Teng, H.Y., *Optimizing Preventive Maintenance for Mechanical Components using Genetic Algorithms*, Reliability Engineering and System Safety, 2001, Vol. 74 (1), pp. 89-97.

[16]   Yang, J., Sung, T. and Jin, Y., *Optimization of the Surveillance Test Interval of the Safety Systems at the Plant Level*, Nuclear Technology, 2000, Vol. 132 (3), pp. 352-365.

[17]   Charles, E. and Kondo, A., *Availability Allocation to Repairable Systems with Genetic Algorithms: a Multi-Objective Formulation*, Reliability Engineering and System Safety, 2003, Vol. 82 (3), pp. 319-330.

[18]   Fonseca, C.M. and Fleming, P.J., *An Overview of Evolutionary Algorithms in Multi-Objective Optimization*, Evolutionary Computation, 1995, Vol. 3 (1), pp. 1-16.

[19]   Holland, J.H., *Adaptation in Natural and Artificial Systems*, Ann Arbor, MI: University of Michigan Press, 1975.

[20]   Goldberg, D.E., *Genetic Algorithms in Search, Optimization, and Machine Learning*, Addison-Wesley Publishing Company, 1989.

[21]   Marseguerra, M. and Zio, E., *Genetic Algorithms: Theory and Applications in the Safety Domain*, In: *The Abdus Salam*

International Centre for Theoretical Physics: Nuclear Reaction Data and Nuclear Reactors, Paver, N., Herman, M. and Gandini, A. Eds., World Scientific Publisher, 2001, pp. 655-695.

[22] Eisinger, S. and Zio, E., *Embedding Local Search Procedures into Genetic Algorithms for the Optimisation of Industrial Systems*, ESREL 2004/ Probabilistic Safety Assessment and Management PSAM VII, 2004, June 14-18, Berlin, Germany, pp.170-178.

[23] Michalewicz, Z., *Genetic Algorithms + Data Structures = Evolution Programs*, Springer Verlag, Berlin, Heidelberg, New York, 1992.

[24] Schwefel, H.-P., *Numerical Optimization of Computer Models*, Ed. John Wiley & Sons, Chichester, 1981.

[25] Sawaragi, Y., Nakayama, H. and Tanino, T., *Theory of Multi-Objective Optimization*, Academic Press, Orlando, Florida, 1985.

[26] Rubinstein, R., Levitin, G., Lisniaski, A. and Ben-Haim, H., *Redundancy Optimisation of Static Series-Parallel Reliability Models under Uncertainty*, IEEE Transactions on Reliability, 1997, Vol. 46 (4), pp. 503-511.

[27] Chern, M.S., *On the Computational Complexity of Reliability Redundancy Allocation in a Series System*, Operations Research Letters, 1992, Vol. 11, pp. 309-315.

[28] Tillman, F.A., Hwang, C.L. and Kuo, W., *Optimization Techniques for System Reliability with Redundancy: a Review*, IEEE Transactions on Reliability, 1977, Vol. R-26, pp. 148-155.

[29] Tillman, F.A., Hwang, C.L. and Kuo, W., *Optimization of System Reliability*, Marcel Dekker, 1980.

[30] Painton, L. and Campbell, J., *Genetic Algorithm in Optimization of System Reliability*, IEEE Transactions on Reliability, 1995, Vol. 44 (2), pp. 172-178.

[31] Coit, D. and Smith, A., *Reliability Optimization of Series-Parallel Systems using Genetic Algorithm*, IEEE Transaction on Reliability, 1996, Vol. 45 (2), pp. 254-266.

[32] Levitin, G., Lisnianski, A. and Elmakis, D., *Structure Optimization of Power System with Different Redundant Elements*, Electric Power Systems Research, 1997, Vol. 43, pp. 19-27.

[33] Levitin, G., Lisnianski, A., Ben-Haim, H. and Elmakis, D., *Redundancy Optimization for Series-Parallel Multi-state Systems*, IEEE Transactions on Reliability, 1998, Vol. 47 (2), pp. 165-172.

[34] Yang, J.-E., Hwang, M.-J., Sung, T.-Y. and Jin, Y., *Application of Genetic Algorithm for Reliability Allocation in Nuclear Power*

Plants, Reliability Engineering and System Safety, 1999, Vol. 65, pp. 229-238.

[35] Gen, M. and Kim, J., *GA-based Reliability Design: State-of-the-Art Survey*, Computer ind. Engineering, 1999, Vol. 37 (1/2), pp. 151-155.

[36] Andrews, J.D. and Bartlett, L.M., *Genetic Algorithm Optimization of a Firewater Deluge System*, Quality and Reliability Engineering International, 2003, Vol. 19, pp. 39-52.

[37] Kiu, S. and McAllister, D.F., *Reliability Optimization of Computer-Communication Networks*, IEEE Transactions on Reliability, 1988, Vol. 37 (5), pp. 475-483.

[38] Kumar, A., Pathak, R., Gupta, Y. and Parsaei H., *A Genetic Algorithm for Distributed System Topology Design*, Computer ind. Engineering, 1995, Vol. 28 (3), pp.659-670.

[39] Levitin, G., Mazal-Tov, Sh. and Elmakis, D., *Genetic Algorithm for Open-Loop Distribution System Design*, Electric Power Systems Research, 1995, Vol. 32, pp. 81-87.

[40] Altiparmak, F., Dengiz, B. and Smith., A., *Reliability Optimization of Computer Communication Network using Genetic Algorithms*, Proceeding of IEEE International Conference on Systems, Man and Cybernetics 5, 1998, Piscataway, NJ, USA, 98CB36218, pp. 4676-4680.

[41] Dhingra, A., *Optimal Apportionment of Reliability & Redundancy in Series Systems under Multiple Objectives*, IEEE Transaction on Reliability, 1992, Vol. 41 (4), pp. 576-582.

[42] Misra, K. and Sharma, U., *An Effective Approach for Multiple Criteria Redundancy Optimization Problems*, Microelectronics and Reliability, 1991, Vol. 31 (2/3), pp. 303-321.

[43] Misra, K. and Sharma, U., *Multi-Criteria Optimization for Combined Reliability and Redundancy Allocation in Systems Employing Mixed Redundancies*, Microelectronics and Reliability, 1991, Vol. 31 (2/3), pp. 323-335.

[44] Yangping, Z., Bingquan, Z. and Dongxin, W., *Application of Genetic Algorithm to Fault Diagnosis in Nuclear Power Plants*, Reliability Engineering and System Safety, 2000, Vol. 67, pp. 153-160.

[45] Wen, F. and Han, Zh., *Fault Section Estimation in Power Systems using a Genetic Algorithm*, Electric Power Systems Research, 1995, Vol. 34, pp. 165-172.

[46] Wen, F. and Chang, C., *A New Approach to Fault Diagnosis in Electrical Distribution Networks using a Genetic Algorithm*, Artificial Intelligence in Engineering, 1998, Vol. 12, pp. 69-80.

[47] Munoz, A., Martorell, S. and Serradell, V., *Genetic Algorithms in Optimizing Surveillance and Maintenance of Components*, Reliability Engineering and System Safety, 1997, Vol. 57 (2), pp. 107-120.

[48] Savic, D., Walters, G. and Knezevi, J., *Optimal Opportunistic Maintenance Policy using Genetic Algorithms, 1:Formulation*, Journal of Quality in Maintenance Engineering, 1995, Vol. 2, pp. 34-49.

[49] Usher, J., Kamal, A. and Syed, W., *Cost Optimal Maintenance and Replacement Scheduling*, IEEE Transactions, 1998, Vol. 30, pp. 1121-1128.

[50] Martorell, S., Carlos, S., Sanchez, A. and Serradell, V., *Constrained Optimization of Test Intervals using a Steady-State Genetic Algorithm*, Reliability Engineering and System Safety, 2000, Vol. 67 (3), pp 215-232.

[51] Marseguerra, M., Zio, E. and Podofillini, L., *Condition-based Maintenance Optimization by means of Genetic Algorithms and Monte Carlo Simulation*, Reliability Engineering and System Safety, 2002, Vol. 77, pp 151-165.

[52] Marseguerra, M. and Zio, E., *Genetic Algorithms for Estimating Effective Parameters in a Lumped Reactor Model for Reactivity Predictions*, Nuclear Science and Engineering, 2001, Vol. 139, pp. 96-104.

[53] Carlos, S., Ginestar, D., Martorell, S. and Serradell, V., *Parameter Estimation in Thermal-Hydraulic Models using the Multidirectional Search Method*, Annals of Nuclear Energy, 2003, Vol. 30, pp. 133-158.

[54] Aldemir, T., Torri, G., Marseguerra, M., Zio, E. and Borkowski, J.A., *Using Point Reactor Models and Genetic Algorithms for On-Line Global Xenon Estimation in Nuclear Reactors*, Nuclear Technology, 2003, Vol. 143 (3), pp. 247-255.

[55] Devooght, J. and Smidts, C., *Probabilistic Reactor Dynamics I. The Theory of Continuous Event Trees*, Nuclear Science and Engineering, 1992, Vol. 111 (3), pp. 229-240.

[56] Siu, N., *Risk Assessment for Dynamic Systems: an Overview*, Reliability Engineering and System Safety, 1994, Vol. 43, pp. 43-74.

[57] Borgonovo, E., Marseguerra, M. and Zio, E., *A Monte Carlo Methodological Approach to Plant Availability with Maintenance, Ageing and Obsolescence*, Reliability Engineering and System Safety, 2000, Vol. 67 (1), pp. 61-74.

[58] Harunuzzaman, M. and Aldemir, T., *Optimization of Standby Safety System Maintenance Schedules in Nuclear Power Plants*, Nuclear Technology, 1996, Vol. 113, pp. 354-367.

[59] N.R.C. U.S. Nuclear Regulatory commission, *Rates of Initiating Events at United States Nuclear Power Plants: 1987-1995*, NUREG/CR-5750.

[60] Parks, G.T., *Multi-Objective Pressurized Water Reactor Reload Core Design using Genetic Algorithm Search*, Nuclear Science and Engineering, 1997, Vol. 124, pp. 178-187.

[61] ICRP Publication 60, *1990 Recommendations of the International Commission on Radiological Protection*, Annals of the ICRP, 1991, Vol. 21 (1-3), Pergamon Press.

5. DEPENDENT FAILURES

5.1 Introduction

All modern technological systems are highly redundant but still fail because of dependent failures. The modeling of this kind of failures is still a critical issue in PSA (Probabilistic Safety Assessment). This is because dependent failures can defeat redundant protective barriers and thus contribute significantly to risk; quantification of such contribution is thus necessary to avoid gross underestimation of risk.

5.2 General classification

Dependent failures may be classified into two main groups [1]:

i.   *Common Cause Failures*

These are multiple failures which result directly from a common or shared root cause. They are commonly termed common cause failures (CCFs). For example, the root cause may be extreme environmental conditions (fire, flood, earthquake, lightning, etc.), failure of a piece of hardware external to the system, or a human error. Operation and maintenance errors can also be the root cause of multiple failures (carelessness, miscalibrations, erroneous procedures). On the contrary, the root cause is not a failure of another component in the system.

'Multiplicity' of the common cause failure is defined as the number of components that fail due to that common cause.

One of the best-known accidents resulting from a common cause failure is the fire at Browns Ferry nuclear power plant (Alabama, USA, 1975) [2]. Two operators used a candle to check for air leaks between the cable room and one of the reactor buildings, which was kept at a negative air pressure differential. The candle's flame was sucked in along the conduct, and the methane seal, used where the cables penetrate the wall, caught fire. The fire went on to damage 2000 cables, including those of automatic Emergency Shut-Down Systems (ESDs), and all manually operated valves, except for four relief valves. Thanks to the availability of these four valves, it was possible to shut down the reactor and avoid nuclear meltdown. As a feedback from the accident, the cables of the different ESDs were placed in separate conduits, with no combustible filling (e.g. urethane foam) [1].

ii.   *Cascading Failures*

These are multiple failures initiated by the failure of one component in the system (as a sort of chain reaction or domino effect). When several components share a common load, failure of one component may lead to increased load on the remaining ones and, thus, to an increased likelihood of failure.

This situation is common in some electrical blackouts of recent occurrence. Indeed, the electrical power transmission systems are typical examples of complex systems under the threat of cascading failures. For example, most of America and Canada east of the Rocky Mountains is supplied by a single network running at a shared supply frequency. This network includes thousands of generators, tens of thousands of transmission lines and network nodes and about one hundred control centers that monitor and control the network flows. All the electrical components have limits on their currents and voltages. If these limits are exceeded, automatic protection devices or the system operators disconnect the components from the system. On the other hand, components can also fail in the sense of misoperation or damage due to aging, fire, weather, poor maintenance or incorrect design or operating settings. The failure of a component causes a transient in which the power flow is redistributed to other components according to circuit laws and to the above mentioned automatic and manual reconfiguration actions.

The transients and readjustments of the system can be local in effect or can involve components far away, so that a component disconnection or failure can effectively increase the loading of many other components throughout the network, possibly causing a blackout in which the initial trigger event is followed by a cascade of events. Examples of trigger events are short circuits of transmission lines through untrimmed trees, protection device misoperation and bad weather.

The interactions between component failures which may lead to cascading failures are stronger when components are highly loaded, as is the trend in today's liberalized market. For example, if a more highly loaded transmission line fails, it produces a larger transient, there is a larger amount of power to redistribute to other components and failures in nearby protection devices are more likely. Moreover, if the overall system is highly loaded, components have smaller margins so they can tolerate smaller increases in load before failure, the system nonlinearities and dynamical couplings increase and the system operators have fewer options, less safety margins and more stress.

In light of recent events, such as the 2003 Northeast Blackout, and the prevalent dependencies on electric power, it is recognized that a large disruption in the bulk power system, either due to random events or intentional attacks, may result in widespread consequences. These consequences could include economic, social, physical, and psychological impacts. The blackout of the Northeast on August 14, 2003, that affected over 50 million people, has been estimated to have had an economic impact between $4 billion and $10 billion in the United States alone [3].

The analysis of power transmission systems is usually carried out either by deterministic analysis of estimated worst cases or by Monte Carlo simulation (chapter 2) of moderately detailed probabilistic models that capture steady state interactions [4]. Combinations of likely contingencies and some dependencies between failure events, e.g. common mode or common cause, are sometimes considered. These analyses, however, address only the first few likely failures rather than the propagation of rare or unanticipated failures in a cascade leading to a large blackout.

Large cascading failures are present also in other network systems such as the communication, social and economic ones [5]. For example, they take place also on the Internet, when failures requiring a rerouting of the information traffic eventually lead to an avalanche of overloads on routers that are not equipped for significant extra traffic. For instance in October 1986, the redistribution resulted in a congestion regime with a large drop in the network performance, the speed of the connection dropping by a factor of 100 between the Lawrence Berkeley Laboratory and the University of California at Berkeley, two places separated only by 200 m [6,7].

The recent occurrences of cascading failures, of somewhat increasing frequencies, have greatly increased the attention of the service providers, operators and users to small local initial shocks, such as the breakdown of an Internet router or of an electrical substation or line, which can trigger avalanche mechanisms affecting a considerable fraction of the service network, possibly collapsing the system.

5.3 Identification of dependent failures and protection from their occurrence

To identify dependent failures, processes of hazard identification by approaches like the Failure Modes and Effects Analysis (FMEA) must be extended to encompass potential interdependencies between the components which may lead to common cause or cascading failures [8]. Dedicated anlyses must be carried out to identify those components who may share the risk of dependent failures [9].

With respect to failure protection, in general the most important defense against accidental component failures is the use of redundancy. The example of the fire at Brown Ferry has shown, however, that redundancy itself is not enough, precisely because of dependent failures. Some general defensive strategies to specifically avoid dependent failures are [1]:

- Barriers (physical impediments that tend to confine and/or restrict a potentially damaging condition)
- Personnel training (ensure that procedures are followed in all operation conditions)
- Quality control (ensure the product is conforming with the design and its operation and maintenance follow the approved procedures and norms)
- Redundancy
- Preventive maintenance
- Monitoring, testing and inspection (including dedicated tests performed on redundant components following observed failures)
- Diversity (equipment diversity as for manufacturing, functional diversity as for the physical principle of operation).

5.4 Definition of dependent failures

From a probabilistic point of view, two events A and B, whose probabilities of occurrence are equal to $P(A)$ and $P(B)$, respectively, are said to be dependent if the probability of their intersection $P(A \cap B)$ is:

$$P(A \cap B) = P(A|B) \cdot P(B) \neq P(A) \cdot P(B) \qquad (5.1)$$

where $P(A|B)$ is the probability of event A conditioned on the occurrence of event B [8].

With respect to the probabilistic dependence of failure events in components and systems, in practice one distinguishes [1]:

1. Common Cause initiating events (external events)

Events that have the potential for initiating a plant transient and increase the probability of failure in multiple systems. These types of events, e.g. fires, floods, earthquakes, loss of offsite power, deserve a full, dedicated risk analysis, which is not expanded here.

2. Intersystem dependences

A.   Functional dependences (from plant design):
      System 2 functions only if system 1 fails.

B.   Shared-equipment dependences:
      Dependences of multiple systems on the same components, subsystems or auxiliary components (e.g. components in different systems fed from the same electrical bus).

C.   Physicals interactions:
      Failures of some system create extreme environmental stresses, which increase the probability of multiple-system failures (e.g. failure of one system to provide cooling results in excessive temperature which causes the failure of a set of sensors).

D.   Human-interaction dependences
Dependences introduced by human actions, including errors of omission and commission (e.g. an operator turns off a system after failing to correctly diagnose the conditions of a plant).

3. Intercomponent dependences

Events or failure causes that result in dependence among the probabilities of failure for multiple components or subsystems. The same cases A-D of intersystem dependences exist for intercomponent dependences as well.

5.5   Methods for dependent-failure analysis

We can distinguish between:

     i.   *Explicit methods*

Involve the identification and treatment of specific root causes of dependent failures at the system level, in the event and fault-tree logic [8]. Examples are earthquakes, fires and floods, which are treated explicitly as initiating events in the risk analysis.

     ii.   *Implicit methods*

Multiple failure events, for which no clear root cause event can be identified and treated explicitly, can be modeled using implicit, parametric models. In these methods, new reliability parameters are added to account for dependent failures.

5.5.1   Examples of explicit methods

5.5.1.1   Intersystem dependences

Let us first consider the intersystem dependences of the kind listed above, referring to a simple example in which two safety systems S_1 and S_2 are expected to intervene upon the occurrence of an initiating event (IE). The generic event tree representing the accident sequences is given in Fig. 5.1.

Fig. 5.1

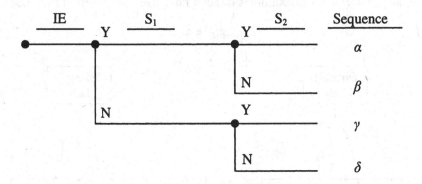

A. Functional dependences

System 2 is not needed (NN) unless system 1 fails. Hence, the event tree of Fig. 5.1 transforms into the one depicted in Fig. 5.2.

Fig. 5.2

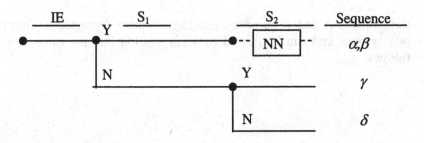

B. Shared-equipment

- Method of the *"event trees with boundary conditions"*

To illustrate the approach for analyzing intersystem dependences of type 2.B., shared-equipment, suppose that the fault trees developed for the systems S_1 and S_2 in Fig. 5.1 contain the same component failures, A and F, as primary events. Examples of such fault trees are given in Fig. 5.3.

Fig. 5.3

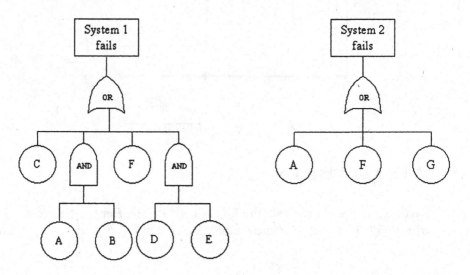

Components A and F introduce shared-equipment dependences into the two systems and can be treated by incorporation into the event tree as follows:

Fig. 5.4

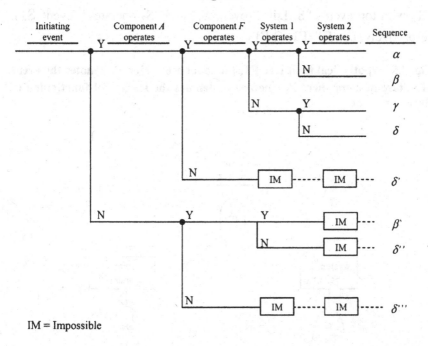

IM = Impossible

To complete the analysis, the system fault trees are quantified as conditional on the states of A and F, which are treated as "house" events [8]. For example, along sequence δ' the fault tree for system S_1 is quantified with $P(A)=1$ and $P(F)=0$, which gives the conditional minimal cut sets {C, B, DE}. On the other hand, along sequence δ the conditions are $P(A)=0$ and $P(F)=0$, which gives the minimal cut sets for system S_1 of {C, DE}.

-   Method of "*Fault tree link*"

The fault trees of systems S_1 and S_2 are linked together, thus developing a single large fault tree for each accident sequence. Considering for example the sequence γ, it can be structured into a fault tree FT_γ with top event "S_1 fails and S_2 operates". Then the fault tree for sequence γ is generated as the intersection ($\cap$) of the two individual fault trees FT_1 and FT_2 with top events "S_1 fails" (event S_1) and "S_2 operates" (event $\overline{S}_2$), respectively, i.e., $FT=FT_1 \cap FT_2$.

Fig. 5.5 Hypothetical fault tree FT_γ for sequence γ. Here X denotes the event of failure of component X, whereas $\overline{X}$ denotes the successful functioning of the component

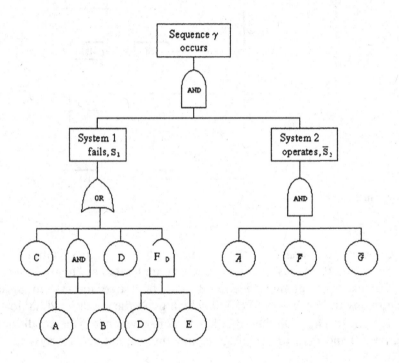

The minimal cut sets can be found as follows [8]:

$$\gamma = S_1 \cap \overline{S}_2 = \left[(A \cap B) \cup C \cup (D \cap E) \cup F \right] \cap \left[\overline{A} \cap \overline{F} \cap \overline{G} \right] =$$

$$= \left[C \cup (D \cap E) \right] \cap \left(\overline{A} \cap \overline{F} \cap \overline{G} \right) \qquad (5.2)$$

$$mcs = \left\{ \overline{AFG}C; \overline{AFG}DE \right\}$$

When rigorously followed, both methods of event trees with boundary conditions and fault tree linking correctly model the shared-equipment dependencies and both entail, apparently, the same level of data processing. However, whereas in the former it is the analyst who must recognize the shared-equipment dependence and explicit it into the event tree, in the latter such dependence is automatically accounted for in the *mcs* of the intersected fault trees representing the sequences.

C. Physical Interactions

System S_2 can operate only if system S_1 operates successfully. When system S_1 fails a physical interaction takes place, which inhibits system S_2. This dependence is represented in the event tree as shown in Fig. 5.6 (IM=Impossible).

Fig. 5.6

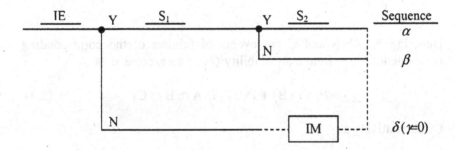

5.5.1.2 Intercomponent dependences

With respect to the intercomponent dependences, let us consider their effects with reference to the three-component system in Fig. 5.7 [1].

Fig. 5.7

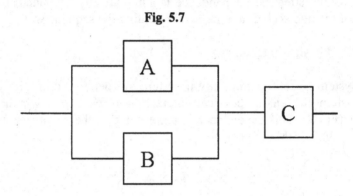

Denoting by A, B and C the events of failures of the corresponding components, the system unavailability Q can be expressed as

$$Q = P(A \cap B) + P(C) - P(A \cap B \cap C) \tag{5.3}$$

Or alternatively as

$$Q = P(A)\, P(B|A)\, [1 - P(C|A \cap B)] + P(C) \quad \textit{for } P(B|A) >> P(B) \tag{5.4}$$

Any cause of failure that affects any pair, or all three, components at the same time will have an effect on the system unavailability. The causes of single and multi-component failures can be represented in the format of a fault tree where the causes appear at the level below the basic component-failure modes and lead to different minimal cut sets (*mcs*), as shown in Fig. 5.8 and Table 5.1. Neglecting these causes of dependent failures (i.e., assuming independence in the component unavailabilities) leads to:

- Optimistic predictions of system availability for components in the same *mcs* (i.e., in parallel)
- Conservative predictions of system availabilty for components in different *mcs* (i.e. in series).

This is shown in the numerical Example 5.1 below [1].

Fig. 5.8 Fault tree for a three-component system with independent and dependent causes of failures

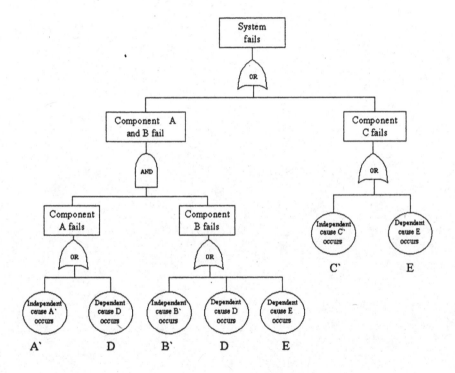

D = common cause of failure shared by redundant components A and B ($\in$ same *mcs*)

E = common cause of failure shared by components in series ($\notin$ same *mcs*)

Table 5.1

Minimal cut sets	
Without common causes of failures	With common causes of failures
$A \cap B$	$A` \cap B`$
C	C`
	D
	E

Example 5.1 [1]:

The unavailability of the system in Fig. 5.7 under different assumptions of common causes of dependent failures is reported in Table 5.2.

Table 5.2

	Fault-tree quantification case	
	A, B (parallel configuration ≡ *mcs*)	
Parameter	Case 1 No common cause, no single failures	Case 2 Common causes shared by components A and B, no single failures
$P(A')$	$1.0 \cdot 10^{-3}$	$9.9 \cdot 10^{-4}$
$P(B')$	$1.0 \cdot 10^{-3}$	$9.9 \cdot 10^{-4}$
$P(C')$	0	0
$P(D)$	0	$1.0 \cdot 10^{-5}$
$P(E)$	0	
Q	$1.0 \cdot 10^{-6}$	$1.1 \cdot 10^{-5}$
Comment	Effects of a common cause shared by components in parallel configuration (i.e., ∈ same *mcs*). Unavailability is 10^{-3} for both components. As the common cause contribution $P(D)$ to the component failure probability is varied from 0 to 1% of the component failure probability ($P(A')=P(B')= 10^{-3}$), the system unavailability Q is <u>increased</u> by more than a factor of 10.	

Parameter	B, C (series configuration ≡ different *mcs*)	
	Case 3 No redundancy, no common cause failure	Case 3 No redundancy, no common cause failure
$P(A')$	1	1
$P(B')$	$1.0 \cdot 10^{-3}$	$5.0 \cdot 10^{-4}$
$P(C')$	$1.0 \cdot 10^{-3}$	$5.0 \cdot 10^{-4}$
$P(D)$	0	0
$P(E)$	0	$5.0 \cdot 10^{-4}$
Q	$2.0 \cdot 10^{-3}$	$1.5 \cdot 10^{-3}$
Comment	Effects of a common cause shared by components in series configuration (i.e., ∉ same *mcs*). Unavailability is 10^{-3} for both components. As the common cause contribution $P(E)$ to the component failure probability increases from 0 to 5% of the component failure probability $(P(B')=P(C')= 10^{-3})$, the system unavailability decreases by 30%. Hence, this type of common cause can usually be conservatively ignored with a small error on the system unavailability.	

The table header reads: **Fault-tree quantification case**

5.5.2 An example of an implicit method for modeling dependent failures

As shown above, in many situations a quantitative failure analysis based on the assumption of independence will lead to unrealistic results. Thus, significant efforts have been made to develop suitable models that quantitatively take into account different types of dependence. One such early example is the square root method adopted in the famous Reactor Safety Study [10].

In the Reactor Safety Study [10], a simple bounding technique was used to estimate the effects of common cause failures on a system. Let us consider a parallel system of two components A and B. The probability that the system is down is $P(A \cap B)$. Since $A \cap B \subseteq A$,

$$P(A \cap B) \leq P(A) \tag{5.5}$$

Similarly,

$$P(A \cap B) \leq P(B) \tag{5.6}$$

Hence,

$$P(A \cap B) \leq min[P(A), P(B)] \tag{5.7}$$

If A and B are independent,

$$P(A \cap B) = P(A)P(B) \tag{5.8}$$

Whereas, if they are positively dependent

$$P(A|B) \geq P(A) \tag{5.9}$$

Thus,

$$P(A \cap B) = P(A|B)P(B) \geq P(A)P(B) \tag{5.10}$$

Combining (5.7) and (5.10),

$$P(A)P(B) \leq P(A \cap B) \leq min[P(A), P(B)] \qquad (5.11)$$

$$\underbrace{}_{P_L} \qquad \underbrace{}_{P_U}$$

$P(A \cap B)$ can be estimated as the geometric mean of P_L and P_U [10]:

$$P_M(A \cap B) \approx \sqrt{P_L P_U} \qquad (5.12)$$

Note that there is no proven theoretical foundation for the choice of the geometric average.

Example 5.2:

As an example of the effects of applying the square root method, consider a system made up of n identical components in parallel configuration. Each component has unavailability q at a specified time t of interest. Let A_i denote the event that component i is down at such time t. Then, $P(A_i)=q$, $i=1,2,...,n$. The application of the square root method requires the calculation of the following quantity:

$$P_L = \prod_{i=1}^{n} P(A_i) = q^n \tag{5.13}$$

$$P_U = \min\left[P(A_1), P(A_2),..., P(A_n) \right] = q \tag{5.14}$$

$$Q_M = \sqrt{P_L P_U} = q^{\frac{n+1}{2}} \tag{5.15}$$

Table 5.3 reports the results of the analysis for the case $q=10^{-2}$ and varying values of the number n of components in the parallel system. Note how the difference in the system unavailability under the dependence and independence assumptions increases as the number of components n increases.

Table 5.3

n	Independent components $(Q=q^n)$	Square root method $Q_M = q^{\frac{n+1}{2}}$
1	10^{-2}	10^{-2}
2	10^{-4}	10^{-3}
3	10^{-6}	10^{-4}
4	10^{-8}	10^{-5}
5	10^{-10}	10^{-6}

5.6 A methodological framework for common cause failures analysis

The treatment of common-cause failures within a probabilistic safety assessment requires four phases [9]:

    i.   System logic model development
    ii. Identification of common-cause component groups
    iii. Common-cause modeling and data analysis
    iv. System quantification and interpretation of results

5.6.1 System logic model development

This phase consists of the following three steps:

1. System familiarization
2. Problem definition (root causes of common failures to be included in the analysis)
3. Logic model development

It is a fundamental phase for any system analysis; with respect to the usual analyses of complex systems more emphasis is put on the identification and representation of dependent failures. The aim of this phase is to identify and understand the physical and functional links in the system, the functional dependences and interfaces and to develop the corresponding logic models of the system (fault trees and event trees), which include the proper representation of the identified dependences.

5.6.2 Identification of common cause component groups

The objectives of this phase are:
    - Identifying group of components potentially involved in dependent failures and thus to be included in the CCF analysis

- Prioritizing the groups for the best resource allocation of the successive analysis
- Providing engineering arguments for data analysis related to common cause failure events and for the identification of defense alternatives to protect against dependent failures.

The end result of this phase is a definition of components for which common cause failures are to be included in the model and the determination of which root causes and coupling mechanisms should be included in the common cause events for the purpose of quantification.

The following general definition of common cause component group can be assumed [9]:

"A common cause component group is a group of similar or identical components that have a <u>significant</u> likelihood of experiencing a common cause event".

Qualitative and quantitative screenings are performed to discriminate the most important groups of components with common cause failures, so as to keep the analysis to a manageable size.

5.6.2.1 *Qualitative screening*

A system and component analysis is made to identify common attributes and mechanisms of failure that can lead to potential dependent failures. A useful check-list of key attributes which can affect component interdependence, may be the following:

- Similarity of component type
- Similarity of component use
- Similarity of component manufacturer
- Similarity of component internal conditions (pressure, temperature, chemistry)
- Similarity of component boundaries and system interfaces
- Similarity of component location name and/or code
- Similarity of component external environmental conditions (humidity, temperature, pressure)

- Similarity of component initial conditions and operating characteristics (standby, operating)
- Similarity of component testing procedures and characteristics
- Similarity of component maintenance procedures and characteristics.

Based on the identified similarities, the various components in the system may be classified into groups of potential dependence. Practical guidelines to be followed in the assignment of component groups are:

- Identical components providing redundancy in the system should always be assigned to a common cause group
- Diverse redundant components which have piece parts that are identically redundant, should not be assumed fully independent in spite of their diversity
- Susceptibility of a group of components to CCFs not only depends on their degree of similarity but also on the existence/lack of defensive measures (barriers) against CCFs.

5.6.2.2 Quantitative screening

In performing quantitative screening for CCF candidates, one is actually performing a complete quantitative common cause analysis except that a conservative and very simple quantification model is used. The following steps are carried out [9]:

1. The fault trees are modified to explicitly include a single CCF basic event for each component in a common cause group that fails all members of the group (Fig. 5.9).

Fig. 5.9 [7]

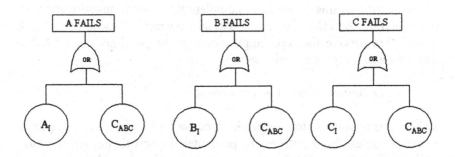

A_I, B_I, C_I = independent failure events
C_{ABC} = common cause failure event which fails all components of the group.

2. The fault trees are solved to obtain the minimal cut sets. To each cut set $\{A_I, B_I, C_I\}$ there is associated also one including C_{ABC}: in the probability-based truncation of minimal cut sets necessary to keep the analysis of large complex systems to a manageable size, the common cause event C_{ABC} may survive because of the relatively large probability, whereas the joint event $\{A_I, B_I, C_I\}$ is often lost, due to its small probability value resulting from the multiplication of the probabilities of the independent failure events A_I, B_I, C_I.

3. Numerical values for the probabilities of the CCF basic events can be estimated by the beta factor model (conservative regardless of the number of components in the CCF basic event) [11]:

$$P(C_{ABC}) = \beta P(A) \qquad \beta = 0.1 \text{ for screening}$$
$P(A)$ = total failure probability in absence of common cause.

4. Those common cause failure events which are found to contribute little to the overall system failure probability are screened out.

5.6.3   Common cause failure modeling and data analysis

The objective of this phase is to complete the system quantification by incorporating the effects of common cause events for those component groups that survive the screening process at the previous stage 5.6.2.2. The following steps are carried out:

1. Definition of common cause basic events

This step is equivalent to a redefinition of the logic model basic events to a level of detail that identifies the particular impacts that common cause events of specified multiplicity may have on the system. An example of this part of the analysis is shown in Fig. 5.10, with reference to a 2-out-of-3 system (the operation of at least two components guarantees system operation or, equivalently, the failure of at least two components leads to system failure).

Fig. 5.10 [9]

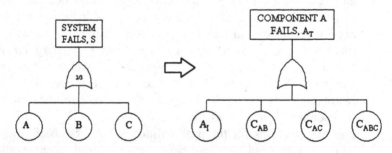

Component level

Common cause impact level
(each component basic event becomes a sub-tree)

$mcs:$ $\{A \cap B\}, \{A \cap C\},$
$\{B \cap C\}$

$mcs = \{A_I \cap B_I\}, \{A_I \cap C_I\}, \{B_I \cap C_I\},$
$\{C_{AB}\}, \{C_{AC}\}, \{C_{BC}\}, \{C_{ABC}\}$

$S = (A \cap B) \cup (A \cap C) \cup (B \cap C)$
$= A{\cdot}B + A{\cdot}C + B{\cdot}C$

$A_T = A_I \cup C_{AB} \cup C_{AC} \cup C_{ABC} =$
$= A_I + C_{AB} + C_{AC} + C_{ABC}$
$S = (A_I \cap B_I) \cup (A_I \cap C_I) \cup (B_I \cap C_I)$
$= A_I{\cdot}B_I + A_I{\cdot}C_I + B_I{\cdot}C_I + C_{AB} + C_{AC} +$
$+ C_{BC} + C_{ABC}$

$X \cup Y =$ union of events X and Y

Note that the basic events of components failures in the left fault tree, related to the system logic without consideration of common cause impacts, are expanded into sub-trees when common causes are considered. This leads to a proliferation in the number of cut sets.

2. *Selection of implicit probability models for common cause basic
   events*

To quantify the probabilities of the common cause failure events, implicit
parametric models are introduced, whose parameters must be identified.
We can distinguish the different models into categories according to the
number of parameters which they rely upon [9]:

- single-parameter models: the beta factor model, which produces
  conservative results for high-redundancy systems
- multi-parameter models: more sophisticated models which
  provide a more realistic assessment of the probabilities of CCF
  events for groups of redundant components of dimension larger
  than two.

Another possible taxonomy divides the components into categories
according to how multiple failures are modeled to occur [1]:

- shock models: the binomial failure rate model which assumes
  that the system is subject to a common cause
  'shock' which occurs at a certain rate; the
  common cause failure frequency is the
  product of the shock rate and the conditional
  probability of failure, given a shock
- non-shock models: direct models use the probabilities of
  common cause events directly, as in the basic
  parameter model; indirect models estimate the
  probabilities of the common cause events
  through the introduction of other parameters.

2.1 The basic parameter model

This non-shock, direct model makes use of the rare-event approximation
for the probability of system failure as the sum of the probabilities of the
minimal cut sets [12]. With reference to the system of Fig. 5.10, its
probability of failure becomes:

$$P(S)=P(A_I)P(B_I)+P(A_I)P(C_I)+P(B_I)P(C_I)+P(C_{AB})+$$
$$+P(C_{AC})+P(C_{BC})+P(C_{ABC}) \tag{5.16}$$

The following assumptions are also made:

i.   The probability of similar events involving similar types of components are the same
ii. The probability of failure of any given basic event within a common cause component group depends only on the number and not on the specific components in that basic event (symmetry assumption).

In the example of Fig. 5.10, this leads to the definition of the following direct probabilities of failure events:

$$\left.\begin{array}{l} P(A_I) = P(B_I) = P(C_I) = Q_1 \\ P(C_{AB}) = P(C_{AC}) = P(C_{BC}) = Q_2 \\ P(C_{ABC}) = Q_3 \end{array}\right\} \begin{array}{l} Q_t = Q_1 + 2Q_2 + Q_3 \\ Q_S = 3Q_1^2 + 3Q_2 + Q_3 \end{array} \tag{5.17}$$

Q_t = total probability of failure of a component

Q_S = probability of failure of the 2-out-of-3 logic system

Generalizing (5.17), the total failure probability of a component in a common cause group of m components is:

$$Q_t = \sum_{k=1}^{m} \binom{m-1}{k-1} Q_k \tag{5.18}$$

Ideally, the Q_k values can be calculated from data. Unfortunately all the necessary data are normally not available. For this reason, other models have been developed that put less stringent requirements on the data. This however is only done at the expense of making additional assumptions that address the incompleteness of the data.

2.2 The Beta-factor model

This single-parameter model is often used for describing dependent failures of two types: inter-component physical interactions (1.C) and human interactions (1.D) [1,11].

The model assumes that Q_t, the total probability of failure of a single component, can be expanded into an independent contribution, Q_I and a dependent one, Q_m, where m is the number of components in the common cause group. This means that the common cause failure is regarded to fail all m components in the group. Hence, within the scheme of the previous basic parameter model, all Q_k's are 0 except Q_I and Q_m:

$$Q_t = Q_I + Q_m \qquad (5.19)$$

A parameter β is introduced as the fraction of the component total failure probability attributable to dependent failures:

$$\beta = \frac{Q_m}{Q_t} = \frac{Q_m}{Q_I + Q_m} \Rightarrow \begin{array}{l} Q_m = \beta Q_t \\ Q_I = (1-\beta)Q_t \end{array} \qquad (5.20)$$

In the case of the 2-out-of-3 system of Fig. 5.10, this leads to the following system failure probability [9]:

$$Q_s = 3(1-\beta)^2 Q_t^2 + \beta Q_t \qquad (5.21)$$

The total component failure probability Q_t and β must be estimated from data.

For systems with more than two units, the beta-factor model does not provide a distinction between different numbers of multiple failures. This simplification can lead to conservative predictions due to the assumption that all units fail when a common-cause failure occurs.

For time-dependent exponential failure probability distributions, assuming constant small rates λ_m and λ_t :

$$\beta = \frac{Q_m(t)}{Q_t(t)} = \frac{1-e^{-\lambda_m t}}{1-e^{-\lambda_t t}} \approx \frac{\lambda_m}{\lambda_t} = \frac{\lambda_m}{\lambda_m + \lambda_t} \tag{5.22}$$

Example 5.3:

Consider a parallel structure of n identical and independent components with failure rate λ. An external event can cause simultaneous failure of all components in the system. Such event may be represented by a hypothetical component C that is in series with the rest of the system.

Let us assume that the system is non repairable and let $R_I(t)$ denote the reliability function of the identical components, while $R_C(t)$ denotes the reliability function of the hypothetical component C.

Then, $R_C(t)=e^{-\beta \lambda t}$ and the reliability of the system is

$$R(t) = \left[1 - \left(1 - R_I(t)\right)^n\right] \cdot R_C(t) = \left[1 - \left(1 - e^{-(1-\beta)\lambda t}\right)^n\right] \cdot e^{-\beta \lambda t} \tag{5.23}$$

Example 5.4:

Consider the three simple systems:
   a) single component
   b) parallel structure of two identical components
   c) 2-out-of-3 system with two identical components

All the components have the same constant failure rate λ. The systems are exposed to common cause failures, which we model through the β factor.

Compute the system reliability function and the MTTF.

Solution

   a) For the single component, common cause failures are irrelevant

$$R_{(a)}(t) = e^{-\lambda t}$$

$$MTTF_{(a)} = \frac{1}{\lambda}$$

(5.24)

b) For a parallel structure of two identical components

$$R_{(b)}(t) = \left[2 \cdot e^{-2(1-\beta)\lambda t} - e^{-2(1-\beta)\lambda t} \right] \cdot e^{-\lambda \beta t} = 2 \cdot e^{-\lambda t} - e^{-(2-\beta)\lambda t}$$

$$MTTF_{(b)} = \frac{2}{\lambda} - \frac{1}{(2-\beta)\lambda}$$

(5.25)

c) For a 2-out-of-3 system with two identical components

$$R_{(c)}(t) = \left[3 \cdot e^{-(1-\beta)\lambda t} - 2 \cdot e^{-3(1-\beta)\lambda t} \right] \cdot e^{-\lambda \beta t} = 3 \cdot e^{-(2-\beta)\lambda t} - 2 \cdot e^{-(3-2\beta)\lambda t}$$

$$MTTF_{(b)} = \frac{3}{(2-\beta)\lambda} - \frac{2}{(3-2\beta)\lambda}$$

(5.26)

Obviously, all three systems have the same *MTTF* when $\beta=1$ (total dependence).

2.3 Multiple Greek letters model

In this model, the following equation allows to compute the probability of common cause failures of order k in a group of multiplicity m [11]:

$$Q_k = \frac{1}{\binom{m-1}{k-1}} \left(\prod_{i=1}^{k} \rho_i \right) (1 - \rho_{k+1}) Q_t \qquad k=1,2,\ldots,m$$

(5.27)

The m-1 parameters are defined as follows:

$\rho_1 = 1$

$\rho_2 = \beta =$ conditional probability of the failure of at least one additional component, given that one has failed

$\rho_3 = \gamma =$ conditional probability of the failure of at least one additional component, given that two have failed

$\rho_4 = \delta =$ conditional probability of the failure of at least one additional component, given that three have failed.

Example 5.5:

Consider the 2-out-of-3 system of Fig. 5.10. Application of the multiple-greek letters model leads to [9]:

$$
\text{2-out-of-3 system} \begin{cases} Q_t = Q_1 + 2Q_2 + Q_3 \\ Q_1 = (1-\beta)Q_t \\ Q_2 = \dfrac{1}{2}\beta(1-\gamma)Q_t \\ Q_3 = \beta\gamma Q_t \end{cases} \tag{5.28}
$$

$$
\Rightarrow \begin{array}{l} Q_S = 2Q_1^2 + 3Q_2 + Q_3 \\ Q_S = 3(1-\beta)^2 Q_t^2 + \dfrac{3}{2}\beta(1-\gamma)Q_t + \beta\gamma Q_t \end{array}
$$

2.4 α-factor model

For the α-factor model, the following equations hold for a common cause failure group of multiplicity m [13]:

$$
\alpha_k^{(m)} = \frac{\dbinom{m}{k} Q_k^{(m)}}{\displaystyle\sum_{k=1}^{m} \dbinom{m}{k} Q_k^{(m)}} \qquad k=1,2,\ldots,m \tag{5.29}
$$

$$\sum_{k=1}^{m} \alpha_k^{(m)} = 1 \quad (normalization) \tag{5.30}$$

$$Q_k^{(m)} = \frac{k}{\binom{m-1}{k-1}} \frac{\alpha_k^{(m)}}{\alpha_t} Q_t \qquad \begin{array}{l} k = 1, 2, ..., m \\[8pt] Q_t = \sum_{k=1}^{m} k\alpha_k \end{array} \tag{5.31}$$

There are m parameters to be estimated, $Q_k^{(m)}$, $k = 1, 2, ..., m$.

Example 5.6:

Consider a common cause failure group of multiplicity $m=3$:

$$\left.\begin{array}{l} \alpha_T = \alpha_1^{(3)} + 2\alpha_2^{(3)} + 3\alpha_3^{(3)} \\[10pt] Q_1^{(3)} = \dfrac{\alpha_1^{(3)}}{\alpha_t} Q_t \\[10pt] Q_2^{(3)} = \dfrac{\alpha_2^{(3)}}{\alpha_t} Q_t \\[10pt] Q_3^{(3)} = \dfrac{3\alpha_3^{(3)}}{\alpha_t} Q_t \end{array}\right\} \begin{array}{l} Q_{S_{(2/3)}} = 3Q_1^2 + 3Q_2 + Q_3 \\[10pt] \quad = 3\left(\dfrac{\alpha_1}{\alpha_t}\right)^2 Q_t^2 + 3\dfrac{\alpha_2}{\alpha_t}Q_t + 3\dfrac{\alpha_3}{\alpha_t}Q_t \end{array} \tag{5.32}$$

2.5 Binomial failure rate (BFR) model

Consider a system composed of m identical components. Each component can fail at random times, independently of each other, with failure rate λ. Furthermore, a common cause shock can hit the system with occurrence rate μ. Whenever a shock occurs, each of the m individual components may fail with probability p, independent of the states of the other components. The term binomial failure rate, which characterizes this model, is used because the number I of individual components failing as a consequence of the shock is binomially distributed with parameters m and p [14]:

$$p[I=i] = \binom{m}{i} p^i (1-p)^{m-i} \quad i=0,1,...,m \qquad (5.33)$$

Two conditions are further assumed:
   i.   Shocks and individual failures occur independently of each other;
   ii.   All failures are immediately discovered and repaired, with negligible repair time.

The assumption that the component will fail independently of each other, given that a shock has occurred, is often not satisfied in practice. The problem can, to some extent, be remedied by defining one fraction of the shocks as being `lethal shocks`, namely shocks that automatically cause all the components to fail ($p=1$). If all the shocks are lethal, one is back to the beta-factor model. Observe that the case $p=1$ corresponds to the situation that there is no built-in protection against these shocks.

The BFR model differs from the beta-factor model in that the former distinguishes between the numbers of multiple-unit failures in a system with more than two units. Indeed, the failure rate of i units in a common cause failure group of multiplicity m is:

$$\lambda_1 = \underbrace{m\lambda}_{} + \underbrace{\mu \left[\binom{m}{1} p(1-p)^{m-1} \right]}_{} \quad Failure\ rate\ of\ i=1\ unit \qquad (5.34)$$

Total Rate of single-unit
contribution due failures from
to independent common cause
failures shocks

$$\lambda_i = \mu \left[\binom{m}{i} p^i (1-p)^{m-i} \right] \quad Failure\ rate\ of\ i\ units,\ i=2,...,m \qquad (5.35)$$

Three parameters need to be estimated in this model, λ, μ and p.

3. Data classification and screening

For the estimation of the parameters of the failure models used in probabilistic risk assessment, the data sources available are typically of two kinds:

- generic raw data
- plant specific data.

For the parameters of a common cause failure model, in principle:

- A complete set of events should be available for each of the common cause component groups.
- A complete set of events should be found for each of the common cause component groups.

In practice, the classification of an event may be done as shown in Tables 5.4 and 5.5 [9]:

Table 5.4 Event classification

Plant (Date)	Status	Event Description	Cause-Effect Diagram
Maine Yankee (August 1977)	Power	Two diesel generators (DG) failed to run due to plugged radiator. The third unit radiator was also plugged. (Ambiguity in the number of components failed in the event)	

Table 5.5 Multiple hypothesis Impact Vector Assessment

Component Group Size	Hypothesis	Hypothesis Probability	I_0	I_1	I_2	I_3	Shock type	Fault Mode
3	H_1	0.9	0	0	1	0	Non lethal	Failure during operation
	H_2	0.1	0	0	0	1		
	Average		P_0	P_1	P_2	P_3		
	Impact Vector ($\bar{I}$)		0	0	0.9	0.1		

As shown in Table 5.5, the impact Vector I of an event that has occurred in a component group of size m has $m+1$ elements, each one representing the number of components that actually failed or could have failed in the event. If k components fail in an event, the k-th element of the impact vector is 1 while all other elements are zero. A condensed representation of the binary impact vector is:

$$I = \{I_0, I_1, ..., I_m\} \tag{5.36}$$

In many reports of dependent failure occurrences, the description of the event is not clear and the classification may require the consideration of several hypotheses H_1, H_2,..., corresponding to different interpretations of the description of the event. The credibility of the different alternative interpretations may be quantified by associating to each one a probability representing the analyst's degree of confidence in that hypothesis. An average non-binary impact vector can then be formed by taking the average of the impact vectors of the various hypotheses. Formally we write:

$$\bar{I} = \{P_0, P_1, ..., P_m\} \tag{5.37}$$

where P_k is the probability that the event involved the failure of k components.

Given a number of impact vectors reporting dependent failures, the number of events in each impact category can be calculated by adding the average impact vectors:

$$n_k = \sum_j P_k(j) \tag{5.38}$$

where n_k is the total number of events involving the failure of k similar components in the group and P_k (j) is the k-th element of the impact vector j.

4. Parameter estimation

The purpose of this phase of the analysis is to use the data available on dependent failures to estimate either the basic event probabilities directly (within the basic parameter model) or the parameters of the common cause failure models (beta factor, BFR, etc.). The information provided by the set of impact vectors derived from recorded data amounts to the number of events in which $1,2,3,\dots m$ components failed.

Example 5.7:

Beta-factor estimation for a two-train redundant standby safety system tested for failures [1].

From the definition (5.20) of β for a common cause group of multiplicity $m=2$, for which the available recorded evidence gives n_1 failures of single components and n_2 failures of both components, one has:

$$\beta = \frac{Q_2}{Q_t} = \frac{Q_2}{Q_1 + Q_2} = \frac{\dfrac{n_2}{N_2}}{\dfrac{n_1}{N} + \dfrac{n_2}{N_2}} \tag{5.39}$$

An estimate of the total single-component failure probability Q_t is assumed to exist. On the contrary, both the number N of single-component demands to start and the number N_2 of tests for common-cause failures are unknown.

The unknown value N of the number of single-component demands to start can be estimated from the known single-component failure probability Q_t as:

$$Q_t = \frac{n_1 + 2n_2}{N} \Rightarrow N = \frac{n_1 + 2n_2}{Q_t} \tag{5.40}$$

The unknown number N_2 of effective tests for common cause failures depends on the surveillance testing strategy:

S_1: Both components are tested at the same time. Then,

$$N_2 = \frac{N}{2} \tag{5.41}$$

$$Q_2 = \frac{n_2}{N/2} \tag{5.42}$$

and the parameter β is estimated from (5.39) as:

$$\beta = \frac{Q_2}{Q_t} = \frac{\dfrac{n_2}{N/2}}{\dfrac{n_1}{N} + \dfrac{n_2}{N/2}} = \frac{2n_2}{n_1 + 2n_2} \tag{5.43}$$

S_2: The components are tested at staggered intervals, say once every two weeks and if there is a failure, the second component is tested immediately. In this case, the number of tests against the common cause is higher because each successful test of a component is a confirmation of the absence of the common cause:

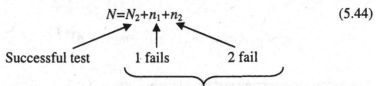

$$N = N_2 + n_1 + n_2 \tag{5.44}$$

Successful test   1 fails   2 fail

These arise because of the failure of the first component, which occurs n_1 times on its own and n_2 times in conjunction.

Then,

$$Q_2 = \frac{n_2}{N_2} \approx \frac{n_2}{N} \quad (n_1 \text{ and } n_2 << N) \tag{5.45}$$

and the parameter β is estimated from (5.39) as:

$$\beta = \frac{Q_2}{Q_t} = \frac{\dfrac{n_2}{N}}{\dfrac{n_1 + n_2}{N}} = \frac{n_2}{n_1 + n_2} \tag{5.46}$$

Estimates of β are therefore based on particular assumptions on the testing strategies.

Example 5.8

Binomial failure rate model parameter estimation [1].

In the BFR model, we need to estimate the two parameters p and μ. The shock rate μ is not directly available from the data because:

- Shocks that do not cause any failure are not observable
- Single failures from common-cause shocks may not be distinguishable from single independent failures.

We define the rate of dependent failures of any multiplicity as:

$$\lambda_+ = \sum_{i=2}^{m} \lambda_i = \mu \left[1 - (1-p)^m - mp(1-p)^{m-1} \right] \tag{5.47}$$

The following experimental data is available:

N_i = number of observations of i concurrent failures

$N_+ = \sum_{i=2}^{m} N_i$ = number of observations of dependent failures of any multiplicity order

The unknown parameters p, μ, λ_1 can be estimated by maximizing the likelihood of the observed data:

$$P_T[N_1 = n_1, N_2 = n_2, ..., N_m = n_m] =$$
$$= P_1[N_1 = n_1] \cdot P_+[N_+ = n_+] \cdot P_m[N_2 = n_2, ..., N_m = n_m] \tag{5.48}$$

For a given observation time T, the variables N_1 and N_+ have Poisson distributions with parameters $\lambda_1 T$ and $\lambda_+ T$, respectively. Maximizing the likelihoods P_1 and P_+ in (5.48) we get:

$$\hat{\lambda}_1 = \frac{n_1}{T} \tag{5.49}$$

$$\hat{\lambda}_+ = \frac{n_+}{T} \tag{5.50}$$

The third factor in the likelihood, P_m, follows a multinomial distribution and the estimate of the value of p which maximizes P_m is found from:

$$S = \frac{mn_+ p\left[1 - (1-p)^{m-1}\right]}{1 - (1-p)^m - mp(1-p)^{m-1}} \tag{5.51}$$

where $S = \sum_{i=2}^{m} i \cdot n_i$ is the total number of units failing in multiple-failure event occurrences. For example, for $m=3$,

$$p = \frac{3(S - 2n_+)}{2S - 3n_+} \tag{5.52}$$

Having found estimates for λ_1, λ_+ and p, one can find an estimate of μ from (6.47).

Example 5.9 [1]

Let us consider the modeling of common cause failures in the Auxiliary Feedwater System (AFWS) of a nuclear Pressurized Water Reactor (PWR) (Fig. 5.11).

Each train of the AFWS is considered as a unit and all of the failure occurrences as failures of the unit to start.

Fig. 5.11 Simplified scheme of an AFWS [1]

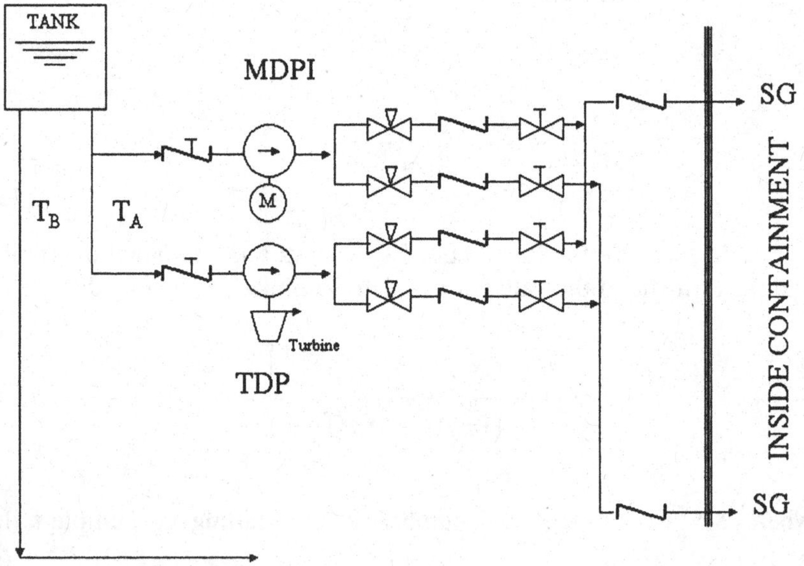

T_A, T_B = Trains A and B
MDP = electrical motor-driven-pump
TDP = steam turbine-driven pump
SG   = steam generator (1 to 4)

Three different types of units (T,M,D) are present in U.S. Nuclear Power Plant AFWS. Some plants have only one unit, whereas others have redundant units either of the same types.

Tables 5.6 and 5.7 report available data of failure occurrences and operating experience on AFWS at U.S. nuclear power plants (at the time of presentation of the example in [1]).

Table 5.6 Instances of multiple failures in PWR auxiliary feedwater systems N_e =11 **multiple failure events for a total of** $Nc=24$ **unit failures**

Plant	Date	Number of failures and failed train type	Number of trains		
			M	T	D
Calvert Cliffs Unit 1	5/76	2/T,T	0	2	0
Haddam Neck	7/76	2/T,T	0	2	0
Kewaunee Unit 1	8/74	2/M,M	2	1	0
	10/75	2/M,T	2	1	0
	11/75	3/M,M,T	2	1	0
Point Beach Unit 1	4/74	2/M,M	2	1	0
Robert F. Ginna	12/73	2/M,M	2	1	0
Trojan Unit	1/76	2/T,D	0	1	1
	12/77	2/T,D	0	1	1
Turkey Point Unit 3	5/74	3/T,T,T	0	3	0
Turkey Point Unit 4	6/73	2/T,T	0	3	0

Table 5.7 Summary of PWR auxiliary feedwater experience

Sum of number of systems times length of service	1874 system-months
Contribution to above by multiple-unit systems	1641 system-months
Contribution to above by 2-units systems	474 system-months
Contribution to above by 3-units systems	1167 system-months
Sum of number of units times length of service	4682 unit-months
Contribution to above by multiple-unit systems	4449 unit-months
Total number of single failures	69
Number of single failures in multiple-unit systems (N_i)	68
Number of multiple-unit failure events (N_e)	11
Number of unit failures in dependent-failure occurrences (N_c)	24

The failure probabilities of 1-out-of-2 and 1-out-of-3 systems can be estimated directly from the data as follows:

System per-demand failure probability:

$$\frac{6 \text{ system failures}}{1641 \text{ system-months of operation}} = 3.7 \cdot 10^{-3} \qquad (5.53)$$

(1-out-of-2) Two-component per-demand failure probability:

$$\frac{4 \text{ system failures}}{474 \text{ system-months of operation}} = 8.4 \cdot 10^{-3} \qquad (5.54)$$

(1-out-of 3) Three-component per-demand failure probability:

$$\frac{2 \text{ system failures}}{1167 \text{ system-months of operation}} = 1.7 \cdot 10^{-3} \qquad (5.55)$$

- *Beta-factor model:*

From the definition (5.20) of the beta factor, an estimate can be found as:

$$\hat{\beta} = \frac{\lambda_m}{\lambda_m + \lambda_i} = \frac{N_c/T}{N_c/T + N_i/T} = \frac{2n_2 + 3n_3}{n_1 + 2n_2 + 3n_3} = \frac{24}{24 + 68} = 0.26$$

$$(5.56)$$

Note that the number of occurrences of multiple-unit failures, N_e, should not be confused with N_c in determining $\hat{\beta}$. A common error is to substitute N_e for N_c.

Assuming one complete (i.e., all units) system demand for each calendar month, the per-demand probability of failure-to-start for a 1-out-of-2 system is:

$$\left.\begin{array}{l}Q_1 = (1-\beta)Q_t \\ Q_2 = \beta Q_t\end{array}\right\} \Rightarrow Q_S = Q_{1-2} = (1-\beta)^2 Q_t^2 + \beta Q_t$$

$$= \underbrace{(1-\beta)^2 \lambda_{tot}^2}_{\text{multiple independent failures}} + \underbrace{\beta \lambda_{tot}}_{\text{common cause failures}} \tag{5.57}$$

For the estimate of the total failure rate,

λ_{tot} = probability of failure on demand of a single unit =

$$= \frac{n_1 + 2n_2 + 3n_3}{N} = \frac{N_i + N_c}{T \cdot 1 (\equiv N)} = \frac{68 + 24}{4682} = 0.02$$

so that the per-demand probability of failure-to-start for the 1-out-of-2 system is:

$$Q_{1-2} = \left[(1-0.26)(0.02)\right]^2 + (0.026)(0.02)$$
$$= 2 \cdot 10^{-4} + 5.2 \cdot 10^{-3} = 5.4 \cdot 10^{-3} \tag{5.58}$$

Note that data from both two- and three-unit systems have been used to obtain a failure probability estimate for a two-unit system.

For a 1-out-of-3-unit system, the contribution from multiple independent failures is negligible, so that the probability of failure to start is:

$$Q_{1-3} = (1-\beta)^3 \lambda_{tot}^3 + \beta \lambda_{tot} = 5.2 \cdot 10^{-3} \tag{5.59}$$

The beta-factor model gives a comparatively higher failure probability for 3-unit systems and a slightly lower probability for 2-unit systems than the values calculated directly from data. Also, note that the system failure probability estimates obtained by the beta-factor model are more robust than those obtained directly from the data.

- *Binomial failure-rate model*

As before, the equation in terms of failure rates λ are converted to failure-to-start probabilities assuming one system demand per calendar month.

$$Q_1 = \frac{N_1}{T \cdot 1} = \frac{68}{1641} = 0.0414$$

where N_1 is the number of observations of single failures

$$Q_2 = \frac{N_t}{T \cdot 1} = \frac{11}{1641} = 0.0067$$

where N_t is the number of observations of multiple failures of any order. Only data from the 7 three-unit systems can be used as evidence for the parameter p (three components group). For this example, the total number of units failing in the 7 multiple-failure occurrences in three-unit systems, S, is 16. Thus,

$$\hat{p} = \frac{3 \cdot (S - 2n_+)}{2S - 3n_+} = \frac{3 \cdot (16 - 2 \cdot 7)}{2 \cdot 16 - 3 \cdot 7} = 0.55 \tag{5.60}$$

$$1 - \hat{p} = 0.45 \tag{5.61}$$

Then, the per-demand common-cause failure shock rate is estimated from (5.47)

$$0.0067 = \mu \left[1 - (0.45)^3 - 3(0.55)(0.45)^2 \right]$$
$$\hat{\mu} = 0.0118 \tag{5.62}$$

Substituting these estimates in (5.35), the per-demand system failure probabilities for 2-out-of-3 units failing is evaluated:

$$Q_{1-2} = (0.0118) \binom{3}{2} (0.55)^2 (0.45)^{3-2} = 4.8 \cdot 10^{-3} \tag{5.63}$$

and for 3-out-of-3 units failing is

$$Q_{1-3} = (0.0118)\binom{3}{3}(0.55)^2(0.45)^{3-3} = 1.9 \cdot 10^{-3} \qquad (5.64)$$

- *Discussion and comparison of Beta-factor and Binomial-Failure-Rate models*

   - The Beta-factor model requires the estimation of 2 parameters: λ, β
   - The Binomial Failure Rate model requires the estimation of 3 parameters: λ, μ, p
   - With both models, one should keep in mind that we are trying to describe the complex reality of multiple failures by means of quite simple models. In certain cases, the models may be inadequate and one should then resort to more complicated (but also more data-demanding) models.
   - In the example, the Beta-factor model estimates the system failure probabilities:

        Q_{1-2} in 2-unit systems

        Q_{1-3} in 3-unit systems

     whereas it does not estimate the probability of 2 components failing in the 1-3 system, $Q_{1-3}^{(2)}$ (but compensates by overestimating Q_{1-3}).
   - The BFR model estimates p from data only taken from three-unit systems and that is why it fits the direct estimate by three units data almost perfectly ($1.9 \cdot 10^{-3}$ against $1.7 \cdot 10^{-3}$).

5.7 References

[1]   NRC Nuclear Regulatory Commission, Probabilistic Risk Analysis: *Procedures Guide, Rep. NUREG/CR-2300*, Washington, DC, 1983.

[2]   http://www.nrc.gov/reading-rm/doc-collections/gen-comm/bulletins/1975/bl75004a.html

[3]   https://reports.energy.gov/B-F-Web-Part1.pdf

[4]   Billington, R., and Allan, R. N., *Reliability Evaluation of Power Systems*, (2nd ed.), Plenum Press, NY, 1996.

[5]   Watts, D.J., Proc. Natl. Acad. Sci. U.S.A., 99, 5766, 2002.

[6]   Jacobson, V., Comput. Commun. Rev. 18, 314, 1988.

[7]   Guimerà, R., Arenas, A., Dìaz-Guilera, A. and Giralt, F., *Dynamical Properties of Model Communication Networks*, Phys.Rev. E 66, 026704, 2002.

[8]   Zio, E., An Introduction to the Basics of Reliability and Risk Analysis, *Series in Quality, Reliability and Engineering Statistics*, Vol.13, World Scientific, Singapore, 2007.

[9]   Mosleh, A., *Common Cause Failures: An Analysis methodology and Examples*, Reliability Engineering and System Safety, 1991, Vol. 34, pp.249-292.

[10]   NRC Nuclear Regulatory Commission, Reactor Safety Study: *An Assessment of Accident Risks in US Commercial Nuclear Power Plants*, Rep. WASH-1400-MR (NUREG-75/014), Washington, DC, 1975.

[11]   Fleming, K. N., *A Reliability Model for Common Mode Failure in Redundant Safety Systems*, In proceedings of the 6th Annual Pittsburgh Conference on Modelling and Simulation, General Atomic report GA-A13284, 23-25 April 1975.

[12]   Vaurio, J. K., *Availability of Redundant Safety Systems with Common Mode and Undetected Failures*, Nuclear Engineering and Design, 1980, Vol. 58, pp. 415-424.

[13]   Mosleh, A. and Siu, N. O., *A Multi-Parameter, Event Based Common-Cause Failure Model*. In proceeding of the 9th International Conference on Structural Mechanics in Reactor Technology, Lausanne, Switzerland, M73, 1987.

[14]   Atwood, C. L., *Common Cause Fault Rates for Pumps*, Report NUREG/CR-1098, 1983.

6. IMPORTANCE MEASURES

6.1 Introduction

From a broad perspective, Importance Measures (IMs) aim at quantifying the contribution of components or basic events to the considered measure of system performance, which, depending on the application, can be the system reliability, unreliability, unavailability or risk. For example, in the nuclear field, the calculation of IMs is a relevant outcome of the Probabilistic Safety Assessment (PSA) of nuclear power plants. In this framework, IMs evaluate the importance of components (or more generally, events) with respect to their impact on a risk measure, usually the Core Damage Frequency or the Large Early Release Frequency. In other system engineering applications, such as aerospace and transportation, the impact of components is considered on the system unreliability or, for renewal systems such as the manufacturing production and power generation ones, on the system unavailability.

Information about the importance of the components constituting a system, with respect to its safety and availability, is of great practical aid to system designers and managers. Indeed, the identification of which components mostly determine the overall system behavior allows one to trace system bottlenecks and provides guidelines for effective actions of system improvement.

IMs were first introduced by Birnbaum [1]. The Birnbaum importance measure gives the contributions to the system reliability due to the reliability of the various system components. Components for which a variation in reliability results in the largest variation of the system reliability have the highest importance. Fussell and Vesely later proposed a measure based on the cut-sets importance [2]. According to the Fussell-Vesely measure, the importance of a component depends on the number and on the order of the cut-sets in which it appears. Other concepts of importance measures have been proposed and used, based on different views of the influence of the components on the system performance. Structural importance measures account for the topographic importance of the logic position of the components in the system [3,4]. Criticality importance measures consider the conditional probability of failure of a component, given that the system has failed [5,6]. Joint importance measures account for the interactions of components in their contribution to system performance [7,8].

Lately, IMs are being widely used in risk-informed applications of the nuclear industry to characterize the importance of basic events, i.e. component failures, human errors, common cause failures, etc., with respect to the risk associated to the system [9-12]. In this framework, two other measures are frequently used: the risk reduction worth and the risk achievement worth [9]. The former one is a measure of the 'worth' of the basic event in achieving the present level of system risk and, when applied to components, it highlights the importance of maintaining the current level of reliability with respect to the basic failure events associated to such components. The latter one, the risk reduction worth, is associated to the maximum decrease in risk consequent to an improvement of the component associated with the basic event considered.

The use of importance measures in risk-informed applications relates to the ranking, or categorization, of components or, more generally, basic events, with respect to their risk-significance or safety-significance. A distinction is made between ranking and categorization [9]: the purpose of ranking is generally to arrange items in order of increasing or decreasing importance; the purpose of categorization is to allocate items into groups, according to some pre-set guidelines or criteria.

Distinction is made also between risk-significance and safety-significance [9,13]. Depending on the application, it may be appropriate to rank or categorize components or basic events, with respect to risk-significance or with respect to safety-significance. Risk-significance and safety-significance are regarded as complementary ways of identifying the role of components or basic events, in determining the risk from operation of the system. On one side, an individual component (basic event) can be identified as being risk-significant if its failure or unavailability (occurrence) contributes significantly to the measures of system risk, e.g.

the core damage frequency, the large early release frequency, the unreliability or the unavailability. Safety-significance, instead, is related to the role that the component or the basic event plays in the prevention of the occurrence of the undesired system end state. In other words, safety significance refers to the significance of a contribution to the probability of system success.

In general, there are two types of applications in risk-informed regulation. The first focuses on the high significance group, with the aim of gaining a reduction in the risk associated to system operation. A categorization according to risk-significance is more appropriate in this case since it focuses on the components that contribute most to the chosen measure of risk.

The second type of applications aims at rendering more effective and less costly the requirements and Operation and Maintenance (O&M) activities by focusing them on what is risk-important, while relaxing those on the low significance group [9], provided that at most only a small risk increase results, well within the limits.

In this chapter, some typical IMs used to characterize the importance of binary components, i.e. components which can be either functioning or faulty, are presented. The definitions of the IMs will be given with respect to the system reliability $R(t)$ and failure probability $F(t) = 1 - R(t)$, which will be taken as the measures of the risk associated to the system. The extension of the definitions of the IMs to multi-state components and systems is illustrated at the end of the chapter.

6.2 Birnbaum's measure

Consider a system of n components. Let $\underline{r}(t) = (r_1(t), r_2(t), ..., r_n(t))$ be the vector of the reliabilities of the individual components at time t and let $R(\underline{r}(t))$ be the system reliability, dependent on the individual components reliabilities and on the system configuration. Birnbaum's measure of importance of the j-th component is defined as:

$$I_j^B(t) = \frac{\partial R(\underline{r}(t))}{\partial r_j(t)} \tag{6.1}$$

$I_j^B(t)$ is a differential sensitivity measure (chapter 7): if $I_j^B(t)$ is large, a small change in the reliability $r_j(t)$ of component j will lead to a large change in the system reliability R at time t.

Properties of Birnbaum's measure are:

    i. $0 \leq I_j^B(t) \leq 1$

    ii. When $R(\underline{r}(t))$ is a linear function of $\underline{r}(t)$ and if all components j are independent, then $I_j^B(t)$ does not depend on $r_j(t)$, $j = 1, 2, ..., n$.

Example 6.1 [6]

Consider three configurations of three binary components characterized by the following reliabilities at a given time: $r_1 = 0.98$, $r_2 = 0.96$, $r_3 = 0.94$. Table 6.1 reports the Birnbaum importances of the three components in the different configurations.

Table 6.1

Configuration #	System configuration (system components)	R	I_1^B	I_2^B	I_3^B
I	Series (1-2)	0.9408	r_2=0.96	r_1=0.98	/
II	Parallel (1-2)	0.9992	1-r_2=0.04	1-r_1=0.02	/
III	2-out-of-3 (1-2-3)	0.9957	0.0952	0.0776	0.0584

With reference to the system configuration I, comprising components 1 and 2 in series, component 2, the less reliable component, is ranked highest with respect to the Birnbaum importance. This result is general: in a series system the Birnbaum measure prioritizes components according to increasing values of reliability. The result is also reasonable from the logic structure of the system: in a series system the reliability is driven by the least reliable component, which constitutes the system bottleneck.

As for configuration II, comprising components 1 and 2 in parallel, the importance measure I_j^B ranks highest component 1, the most reliable one. Again, there is a structural explanation to this ranking: in a parallel system, the system behavior is driven by the most reliable components and I_j^B prioritizes components accordingly. This is confirmed also by the ranking produced by I_j^B for the components of the system in configuration III, where the three components 1, 2 and 3 are in a 2-out-of-3 logic of operation. Again, Birnbaum's importance decreases with decreasing reliability of the components.

6.2.1 Relation with the system structure function

The top event of a fault tree can be represented by an indicator variable X_T which is a Boolean function of the Boolean variables $X_1, X_2, ..., X_n$ describing the states of the n events of the system [14].

$$X_T = \Phi(X_1, X_2, ..., X_n) = \Phi(\underline{X}) \qquad (6.2)$$

Such function is called a structure function and incorporates all the causal relations among the events which lead to the top event. It maps an n-dimensional vector $\underline{X} = (X_1, X_2, ..., X_n)$ of 0's and 1's onto a binary variable equal to 0 or 1.

The structure function can be written expliciting the indicator variable X_j of component j (equal to 1 if the component is functioning and to 0 if failed [14]:

$$
\begin{aligned}
\Phi\left[\underline{X}(t)\right] &= X_j(t)\Phi\left[\underline{X}(t), X_j = 1\right] + \left(1 - X_j(t)\right)\Phi\left[\underline{X}(t), X_j = 0\right] \\
&= X_j(t)\left\{\Phi\left[\underline{X}(t), X_j = 1\right] - \Phi\left[\underline{X}(t), X_j = 0\right]\right\} + \\
&\quad + \Phi\left[\underline{X}(t), X_j = 0\right]
\end{aligned}
\tag{6.3}
$$

By applying the expectation operator $E[\cdot]$ to $\Phi\left[\underline{X}(t)\right]$ and assuming that the components are independent, the system reliability is computed as [14]:

$$
\begin{aligned}
R(\underline{r}(t)) &= r_j(t)\cdot\left\{E\left[\Phi\left[\underline{X}(t), X_j = 1\right]\right] - E\left[\Phi\left[\underline{X}(t), X_j = 0\right]\right]\right\} + \\
&\quad + E\left[\Phi\left[\underline{X}(t), X_j = 0\right]\right] = \\
&= r_j(t)\cdot\left\{R(r_j = 1, \underline{r}(t)) - R(r_j = 0, \underline{r}(t))\right\} + \\
&\quad + R(r_j = 0, \underline{r}(t)) = \\
&= r_j(t)\cdot\left\{R_j^+(t) - R_j^-(t)\right\} + R_j^-(t)
\end{aligned}
\tag{6.4}
$$

where,

-   $R_j^+(t) = R(r_j = 1, \underline{r}(t)) = E\left[\Phi\left[\underline{X}(t), X_j = 1\right]\right]$ is the system reliability when component j is always in its functioning state, i.e. $X_j=1$ and $r_j=1$, throughout the time interval $[0,t]$. It represents the maximum reliability achievement if component j is considered perfect, i.e. always in the functioning state.

- $R_j^-(t) = R(r_j = 0, \underline{r}(t)) = E\left[\Phi\left[\underline{X}(t), X_j = 0\right]\right]$ is the system reliability when component j is always in the faulty state, i.e. $X_j=0$ and $r_j=0$, throughout the time interval $[0, t]$. It represents the maximum reduction in reliability if component j is considered failed with certainty and permanently, or, which is equivalent, removed from the system.

Hence, we can write:

$$I_j^B(t) = \frac{\partial R(\underline{r}(t))}{\partial r_j(t)} = R\left[r_j = 1, \underline{r}(t)\right] - R\left[r_j = 0, \underline{r}(t)\right] = R_j^+(t) - R_j^-(t) \quad (6.5)$$

Since $\Phi\left[\underline{X}(t), X_j = 1\right] - \Phi\left[\underline{X}(t), X_j = 0\right]$ can assume a value equal to 1 or 0 only, then

$$I_j^B(t) = E\left\{\left[\Phi\left[\underline{X}(t), X_j = 1\right] - \Phi\left[\underline{X}(t), X_j = 0\right]\right]\right\}$$

$$= P\left\{\Phi\left[\underline{X}(t), X_j = 1\right] - \Phi\left[\underline{X}(t), X_j = 0\right] = 1\right\}$$

$$= P\{\text{the system state } \left(\underline{X}(t), X_j = 1\right) \text{ is a "critical" path vector}\}$$

(i.e., probability that the system functions only if $X_j=1$, i.e., when component j functions)

Note that the fact that component j is critical tells nothing about the state of component j. The statement concerns only the states of the other system components, $\underline{X}(t)$, i.e. the system must be in such a state that component j being failed ($X_j=0$) leads to the failure of the system and component j being functioning ($X_j=1$) leads to the success of the system.

For example, in the previous series system configuration I, 1 is critical only if 2 is functioning, regardless of the state of component 1 and thus $I_1^B = P[X_2 = 1] = r_2$. Viceversa, for the parallel system configuration II, component 1 is critical only if component 2 is failed and thus $I_2^B = P[X_2 = 0] = 1 - r_2$.

Note also that, denoting by $q_j(t) = 1 - r_j(t)$ the probability that component j fails before t, the Birnbaum importance measure can be defined dually with respect to the system failure probability:

$$I_j^B(t) = \frac{\partial R(\underline{r}(t))}{\partial r_j(t)} = R[r_j = 1, \underline{r}(t)] - R[r_j = 0, \underline{r}(t)] =$$

$$= \frac{\partial F(\underline{q}(t))}{\partial q_j(t)} = F[q_j = 1, \underline{q}(t)] - F[q_j = 0, \underline{q}(t)] = \qquad (6.6)$$

$$= F_j^+(t) - F_j^-(t)$$

where,

- $\underline{q}(t) = (q_1(t), q_2(t), ..., q_n(t))$ is the vector of the unreliabilities at time t of the individual components;
- $F[\underline{q}(t)] = 1 - R[\underline{r}(t)]$ is the system failure probability or unreliability (or, more generally, risk) at time t;
- $F_j^+(t) = F[q_j = 1, \underline{q}(t)] = P\{\Phi[\underline{X}(t), X_j = 0] = 0\}$ is the system failure probability or unreliability when component j is in its faulty state ($X_j=0$) throughout the time interval $[0,t]$. It represents the maximum risk achievement if component j is considered failed with certainty and permanently, or, which is equivalent, removed from the system;
- $F_j^-(t) = F[q_j = 0, \underline{q}(t)] = P\{\Phi[\underline{X}(t), X_j = 1] = 0\}$ is the system failure probability or unreliability when component j remains in the functioning state ($X_j=1$) throughout the time interval $[0,t]$. It represents the maximum reduction in risk if component j is considered perfect, i.e. always in the functioning state.

6.3 Criticality importance

Birnbaum's importance for the component j at time t is independent on the reliability of component j itself, i.e., $I_j^B(t)$ is not a function of $r_j(t)$.

Let $C[\underline{X}(t), X_j = 1]$ be the event that the system is in a state such that j is critical. Such event is independent of the state of j. Then,

$$P\{C[\underline{X}(t), X_j = 1]\} = I_j^B(t) \tag{6.7}$$

The probability that j is critical and ($\cap$) failed at time t is

$$P\{C[\underline{X}(t), X_j = 1] \cap [X_j(t) = 0]\} = I_j^B(t) \cdot [1 - r_j(t)] \tag{6.8}$$

The "criticality importance" $I_j^{cr}(t)$ of component j at time t is defined as the probability that component j is critical for the system and failed at time t, given ($|$) that the system is failed at time t:

$$I_j^{cr}(t) = P\{C[\underline{X}(t), X_j = 1] \cap [X_j(t) = 0] | \Phi[\underline{X}(t)] = 0\}$$

$$= \frac{P\{C[\underline{X}(t), X_j = 1] \cap [X_j(t) = 0]\}}{P\{\Phi[\underline{X}(t)] = 0\}} = \tag{6.9}$$

$$= \frac{I_j^B(t) \cdot [1 - r_j(t)]}{1 - R(\underline{r}(t))} = \frac{I_j^B(t) \cdot q_j(t)}{1 - R(\underline{r}(t))}$$

In other words, $I_j^{cr}(t)$ is the probability that component j has caused the system failure, given that the system is failed at time t. When component j is repaired, the system will start functioning again.

Example 6.2 [6]

Let us consider the same three system configurations as in Example 6.1. Table 6.2 reports the criticality importance measures I_i^{cr} of the three components $i=1,2,3$.

Table 6.2

Configuration #	System configuration (system components)	I_1^{cr}	I_2^{cr}	I_3^{cr}
I	Series (1-2)	$\dfrac{I_1^B(1-r_1)}{1-r_1 r_2} = 0.3243$	$\dfrac{I_2^B(1-r_2)}{1-r_1 r_2} = 0.662$	/
II	Parallel (1-2)	$\dfrac{I_1^B(1-r_1)}{1-r_1-r_2+r_1 r_2} = 1$	1	/
III	2-out-of-3 (1-2-3)	0.4428	0.7219	0.8149

A number of considerations follows:

- The numerator of I_j^{cr} is the probability that the system failure has been caused by component j. Indeed, for instance in case of the series system configuration I, the numerator is $r_2 (1-r_1) = P(\text{component 2 working})\cdot P(\text{component 1 failed})$, i.e. the probability that the system failure has been caused by component 1. Similarly to I^B, in a series system the most important component according to I^{cr} is the least reliable one, for it will most probably be the cause for the system failure.
- In the parallel system configuration II, $I_1^{cr} = I_2^{cr}$ as it should be since if the system is failed, it will start functioning again irrespective of which of the components is repaired. Obviously, this result is general and applies to all simple parallel systems.
- In the 2-out-of-3 system configuration III, the importance is now increasing with decreasing component reliability, opposite to the Birnbaum's measure. Physically, the more unreliable the component, the more its contribution to the system failure.

6.4 Fussell-Vesely importance measure

The Fussell-Vesely importance measure of component j at time t, $I_j^{FV}(t)$, takes into account the fact that a component may contribute to the system failure without being critical. The component contributes to system failure when a minimal cut set (mcs), containing the component, occurs. A cut set is a set $\underline{X}$ such that the structure function $\Phi(\underline{X}) = 1$. A minimal cut set is a cut set that does not have another cut set as a subset. Physically, a minimal cut set is an irreducible cut set: repairing one element of the set, repairs the system [14].

We then define:

$\quad\quad I_j^{FV}(t)$ = probability that at least one mcs containing i is verified

$\quad\quad\quad\quad$ at time t, given that the system is failed at t.

Let:

$\quad\quad m_j$ = number of mcs containing component j, $j=1,2,\ldots,n$

$\quad\quad M_{jh}(t)$ = h-th mcs among those containing component j, verified at time t

$\quad\quad D_j(t)$ = event that at least one mcs that contains component j is verified at time t

$\quad\quad\quad\quad = M_{j1}(t) \cup M_{j2}(t) \cup \ldots \cup M_{jm_j}(t)$, ($\cup$ = logic OR operator)

Then,

$$I_j^{FV}(t) = P\{D_j(t) \mid \Phi[\underline{X}(t)] = 0\}$$
$$= \frac{P\{D_j(t) \cap \Phi[\underline{X}(t)] = 0\}}{P\{\Phi[\underline{X}(t)] = 0\}} \quad\quad (6.10)$$
$$= \frac{P\{D_j(t)\}}{P\{\Phi[\underline{X}(t)] = 0\}}$$

Assuming independent components,

$$P\{\Phi[X(t)]=0\}=1-R(\underline{r}(t))$$
$$P\{M_{jh}(t)\}=\prod_{l\in M_{jh}}(1-r_l(t)) \tag{6.11}$$

However, since component j may belong to several cut sets, the events M_{jh}, $j=1,2,...,n$, $h=1,2,...,m_j$ are not disjoint and independent, even if all the components are independent. Taking the M_{jh}-independence assumption as valid [14],

$$P\{D_j(t)\}\cong 1-\prod_{h=1}^{m_j}\left[1-P\{M_{jh}(t)\}\right] \tag{6.12}$$

and the Fussell-Vesely importance measure may be written as:

$$I_j^{FV}(t)\cong\frac{1-\prod_{h=1}^{m_j}\left[1-P\{M_{jh}(t)\}\right]}{1-R(\underline{r}(t))} \tag{6.13}$$

Adopting the rare-event approximation, neglecting the situation of two or more mcs containing j verified at the same time [14]:

$$I_j^{FV}(t)\cong\frac{\sum_{h=1}^{m_j}P\{M_{jh}(t)\}}{1-R(\underline{r}(t))}=\frac{\sum_{h=1}^{m_j}P\{M_{jh}(t)\}}{F(t)} \tag{6.14}$$

Note from Eq. (6.14) that the numerator of $I_j^{FV}(t)$ can be interpreted as the sum of the terms in the risk equation containing component j, i.e. the fraction of the risk competing to j. Then, $I_j^{FV}(t)$ can be alternatively computed as:

$$I_j^{FV}(t)\cong\frac{F(t)-F_j^-(t)}{F(t)} \tag{6.15}$$

where the numerator actually yields the part of F containing the term q_j.

Example 6.3 [6]

Let us consider the same three system configurations as in Example 6.1. Table 6.3 reports the Fussel-Vesely importance measures I_i^{FV} of the three components $i=1,2,3$.

Table 6.3

Configuration #	System configuration (system components)	I_1^{FV}	I_2^{FV}	I_3^{FV}
I	Series (1-2)	$\dfrac{1-r_1}{1-r_1r_2}=0.3378$	$\dfrac{1-r_2}{1-r_1r_2}=0.6757$	/
II	Parallel (1-2)	1	1	/
III	2-out-of-3 (1-2-3)	0.4651	0.7442	0.8372

A number of considerations follows:

- The values of I_j^{FV}, $j=1,2,\ldots,n$ for the three considered system configurations are very close, if not equal, to the corresponding values of I_j^{cr}, $j=1,2,\ldots,n$. This is no surprise since I_j^{FV} and I_j^{cr} both aim at quantifying the contribution of a component to the system failure probability, though from slightly different perspectives. Compare, for instance, the numerators of the two measures I_1^{FV} and I_1^{cr}, i.e. $(1-r_1)$ and $r_2(1-r_1)$, respectively: according to the Fussell-Vesely measure, component 1 contributes to the system failure when it fails (i.e. when its single-component cut set is verified), independently on whether component 2 is functioning or not; instead, the criticality measure considers the contribution to the system failure of component 1 only when it is this component the responsible of the system failure, i.e. when 1 is failed, but 2 is working.

- In the parallel system configuration II, the system itself constitutes a minimal cut set, that is $D_1(t)=D_2(t)=\left\{\Phi\left[\underline{X}(t)\right]=0\right\}$; it then follows that $I_1^{FV}=I_2^{FV}=1$.

6.5 Risk Achievement Worth and Risk Reduction Worth

With reference to the generic j-th component, two additional importance measures frequently used are the Risk Achievement Worth (RAW) and the Risk Reduction Worth (RRW) [9].

6.5.1 Risk Achievement Worth

The mathematical definition of the RAW of component j at time t is:

$$RAW_j(t) = \frac{F\left[q_j = 1, \underline{q}(t)\right]}{F(t)} = \frac{F_j^+(t)}{F(t)} \tag{6.16}$$

In words, the risk achievement worth is the ratio of the risk when component j is considered always failed in $(0,t)$ ($q_j=1$, $X_j=0$) to the actual value of the risk. It is a measure of the 'worth' of the basic event in achieving the present level of system risk and, when applied to the system components, it highlights the importance of maintaining the current level of reliability with respect to the basic failure event associated to such components. The RAW is a very discriminating measure and it has to be interpreted very carefully. While it can be an appropriate measure for assessing the effect of a temporary change in the component operative condition in which it is made unavailable, if it is used in the context of assessing permanent changes, it is an extreme bounding measure since it considers only complete unavailability as a change.

6.5.2 Risk Reduction Worth

The mathematical definition of the RRW of component j at time t is:

$$RRW_j(t) = \frac{F(t)}{F\left[q_j(t) = 0, \underline{q}(t)\right]} = \frac{F(t)}{F_j^-(t)} \tag{6.17}$$

In words, the risk reduction worth is the ratio of the nominal value of the risk to the risk when component j is always available ($q_j=0$, $X_j=1$). It measures the potential of component j in reducing the risk, by considering the maximum decrease in risk achievable when the component j is always perfectly operating. This measure is useful for identifying improvements which can most reduce risk.

Example 6.4

Let us consider the same three system configurations as in Example 6.1. Tables 6.4 and 6.5 report the RAW and the RRW of the three components $i=1,2,3$, respectively.

Table 6.4

Configuration #	System configuration (system components)	RAW_1	RAW_2	RAW_3
I	Series (1-2)	$\dfrac{1}{q_1+q_2-q_1q_2}=16.89$	$\dfrac{1}{q_1+q_2-q_1q_2}=16.89$	/
II	Parallel (1-2)	$\dfrac{q_2}{q_1q_2}=\dfrac{1}{q_1}=50$	$\dfrac{1}{q_2}=25$	/
III	2-out-of-3 (1-2-3)	22.67	18.31	13.75

Table 6.5

Configuration #	System configuration (system components)	RRW_1	RRW_2	RRW_3
I	Series (1-2)	$\dfrac{q_1 + q_2 - q_1 q_2}{q_2} = 1.48$	$\dfrac{q_1 + q_2 - q_1 q_2}{q_1} = 2.96$	/
II	Parallel (1-2)	$\dfrac{q_1 q_2}{0} = \infty$	$\dfrac{q_1 q_2}{0} = \infty$	/
III	2-out-of-3 (1-2-3)	1.79	3.58	5.38

A number of considerations follows:

- Components in series logic have the same value of RAW, e.g. components 1 and 2 in system configuration I. Indeed, the system is failed if any of the series components is failed, so that $F_j^+ = 1$, $j=1,2,\ldots,n$.
- Components in parallel logic are ranked by RAW opposite to their failure probability. In other words, the more reliable components are ranked first, as suggested by the Birnbaum measure. Reasonably, for parallel components the achievement in risk is highest if the most reliable component is taken out of service.
- Components in series logic are ranked by RRW in increasing order of failure probability. Reasonably, for series components the reduction in risk achievable by improving the component to perfection is highest for the components which contribute most to the system failure, i.e, the least reliable, bottleneck components.
- Components in parallel logic have the same RRW. Indeed, a simple parallel system cannot fail if any of its constituting components cannot fail so that $F_j^- = 0$ and $RRW_j = \infty$, $j=1,2,\ldots,n$.

6.6 Observations and limitations of importance measures

Different importance measures may lead to different rankings of the components. The analyst should be aware of such differences for a proper use of the informative content provided by the measures. Further, all measures are time-dependent: at different times one may get different rankings of the importance of the components.

For a better understanding of risk importance measures, the risk metric F can be represented by the following linear equation [13], which can be derived proceeding as for the dual equation (6.4) for the system reliability:

$$F = \alpha_j \cdot q_j + \beta_j \qquad (6.18)$$

where q_j is the unavailability of the generic component j, $\alpha_j = F_j^+ - F_j^-$ is the coefficient with which q_j appears in the risk equation and $\beta_j = F_j^-$ represents the collection of all the other terms of F that do not contain q_j. Eq. (6.18) holds when component j is independent from the other components. All of the importance measures introduced in the previous Sections can be obtained from the knowledge of q_j, α_j and β_j.

A first issue that can be addressed concerns which IMs are more appropriate to rank or categorize components with respect to risk-significance or with respect to safety-significance (as defined in Section 6.1).

The Fussell-Vesely importance measure is often used as a measure of risk-significance. Using Eqs. (6.18) and (6.15) one can rewrite the expression for I^{FV} as [13]:

$$I_j^{FV} = \frac{\alpha_j q_j + \beta_j - \beta_j}{\alpha_j q_j + \beta_j} = \frac{\alpha_j q_j}{\alpha_j q_j + \beta_j} \approx \frac{\alpha_j q_j}{\beta_j}, \qquad \text{when } \alpha_j q_j << \beta_j \quad (6.19)$$

The assumption $\alpha_j q_j << \beta_j$ is verified in high-risk installations, such as the nuclear ones, which are designed with significant redundancy according to the defense-in-depth principle, so that it is very unlikely that a single component alone contributes much to the risk. I_j^{FV}, then, turns out to be proportional to the unavailability of component j and represents the contribution of component j to the risk metric F. In this sense, I_j^{FV} represents the risk-significance of component j. Note that $\alpha_j q_j$ is the probability of the union of all the minimal cutsets containing component j, so that I_j^{FV} can be alternatively interpreted as the relative contribution to risk of all the minimal cutsets containing component j.

The risk achievement worth RAW is typically used to characterize components according to their safety-significance. By definition, it is a measure of the impact of setting to one the unavailability of the particular component, i.e. of removing it. Using Eq. (6.18), the expression of RAW of (6.16) can be written as:

$$RAW_j = \frac{\alpha_j + \beta_j}{\alpha_j q_j + \beta_j} \approx \frac{\alpha_j}{\beta_j} + 1, \quad \text{when } \alpha_j q_j << \beta_j \tag{6.20}$$

Thus, when $\alpha_j q_j << \beta_j$, RAW_j is independent on q_j and represents the degree of defence against failure provided by the rest of the installation. A high value of RAW_j means that component j is highly safety-significant since the increase in risk due to the unavailability of the component is high. As defined, RAW_j represents an extreme measure of the amplification of risk due to component j, which assumes its complete unavailability. Hence, its use to rank components and define changes in the technical specifications (surveillance and/or test frequencies, etc.) must be very careful: there are very few components, if any, for which the impact of a proposed change is to render them totally ineffective. For this reason, the use of RAW as a safety-significance measure is still controversial and a debate is ongoing among the practitioners on whether other measures could be more suitable.

A most important area of application of importance measures is in support to the establishment of test and maintenance programs [11], which greatly influence the unavailability of components. In this respect, the ranking produced by the Birnbaum importance measure I^B seems to be the most appropriate one. On the other hand, the question on 'what will be the result in terms of risk, when a certain component is taken out of service' is best addressed by RAW.

The other traditional area of application of IMs is in the design of systems and plants. Significant components are selected with the aid of IMs and improvement in the design are introduced for decreasing the unavailability of the selected components and improving the defence-in-depth against their failures. Two IMs are often used for these purposes: I^{FV} and I^B. The I^{FV} importance is used for the selection of components candidate for improvement because contributing to risk the most; then, the information from the Birnbaum importance measure I^B allows identifying for which components the improvements are more effective.

The I^{FV} importance measure is also the most appropriate one for identifying the components that most probably are the cause of system failure and, therefore, it can be used to set up a repair priority checklist.

Though, as above described, IMs are widely used in practice, some shortcomings have been identified that could potentially limit their usefulness. The main issues raised are [9]:

1. IMs produce risk rankings that are not necessarily related to the risk change that results from credible changes to the contributor probabilities q_j. Indeed, IMs deal with changes in reliability or risk only at the extremes (0,1) of the defined range of probability.
2. IMs rank only individual components or basic events whereas they are not directly applicable to combinations or groups of components or basic events. Indeed, there is no simple relationship between the importance measures evaluated at the single component or basic event level and those evaluated at the level of a group of components or basic events. In practice, different basic events may represent different modes of failure or unavailability of a single component and in order to determine the importance of such component one has to consider all the related basic events as a group. Furthermore, many risk-informed applications deal with evaluating the risk change associated to changes in the plant technical specifications (surveillance and/or test frequencies, etc.) which impact a group of components.

3. IMs do not typically consider the credible uncertainty range of components' unavailabilities or basic event probabilities and this raises a doubt on the robustness of the conclusions drawn from importance analyses.

4. IMs have been mainly applied to systems made up of binary components (i.e., components that can be in two states: functioning or faulty). This kind of systems has many practical applications. Yet the hypothesis of dichotomizing the components and system states is often over-simplified and insufficient for describing the real functioning of many systems, whose performance can settle on different levels (e.g. 100%, 80%, 50% of the nominal capacity), depending on the operative conditions of the constitutive multi-state components.

Significant efforts are currently being performed to overcome the above limitations. For example, for the first issue, the fact that the importance measures deal with changes in the probabilities of the basic events only at the extrema 0 or 1 of their ranges, a generalized risk importance measure has been proposed in the literature which depends also on the actual value of a proposed change in the probability of the basic failure event [9].

Then, a Differential Importance Measure, DIM, has been recently introduced to partially overcome the second issue [12]. The DIM is a first-order sensitivity measure (chapter 7) that ranks the parameters of the risk model according to the fraction of the total change in risk due to a small change in the parameters' values, taken one at a time. The DIM bears the important property of additivity: the DIM of a group of components or basic events is the sum of the DIMs of the single components or basic events of the group. The DIM will be addressed in details in the later Section 6.10.

The need for importance measures capable of considering combinations of components arises also when planning a budget-constrained improvement in the reliability of a system design, for example by replacing one of its components with a better-performing one or by inspecting and maintaining it more frequently. Due to the budget constraints, the improvement may need to be accompanied by the sacrifice of the performance of another, less important component. The interactions of these coupled changes to system design must be accounted for when assessing the importance of the system components. To this aim, second order sensitivity measures such as the Joint Reliability Importance (JRI) and Joint Failure Importance (JFI) measures have been introduced [7,8].

A second-order extension of the DIM, named DIM^{II}, has been proposed for accounting for the interactions of pairs of components when evaluating the change in system performance due to changes of the reliability parameters of the components. The extension aims at supplementing the first-order information provided by DIM with the second-order information provided by JRI and JFI for use in risk-informed decision-making [17].

As for the third issue, in general the different importance measures can be considered as random variables. For example, the definition of the risk reduction worth RRW_j is given by Eq. (6.17) as the ratio of the nominal value of the risk F to the risk when component j is always available, F_j^-. Thus, in the most general sense, F and F_j^- are to be considered as random variables characterized by given probability distributions. Then, RRW_j is a random variable for which specific statistics can be calculated, e.g., mean, median, etc. The probability distribution for RRW_j describes the random variability due to the intrinsic randomness of the system performance states. Uncertainties in the parameters (epistemic uncertainties) have also to be included to give the total probability distribution. When the components importance measures are treated as random variables, the non-trivial inter-comparison of their distribution for component ranking must be carried out. Methods have been proposed in the literature, but their application in practice is rare or non-existent, mainly due to the computational difficulty and burden [18,19].

Finally, as for the last issue, the current research activity in the generalization of IMs for application to multi-state systems made up of multi-state components will be illustrated in detail in Section 6.11.

6.7 Generalized risk importance measure

To overcome the fact that the importance measures deal with changes in the probabilities of the basic events only at the extremes 0 or 1 of their ranges, a generalized risk importance measure has been defined by considering the following relative change in risk due to a change in the probability of the basic failure event j from the value q_j to the value q_j^n [9]:

$$\frac{\Delta F_j}{F} = \frac{F_j^n - F}{F} = \left\{ F\left[q_j = 1, \underline{q}(t)\right] - F\left[q_j = 0, \underline{q}(t)\right] \right\} \left(\frac{q_j^n - q_j}{F} \right) \quad (6.21)$$

where:

q_j^n = the considered new value for the probability of the basic failure event j

F_j^n = the system risk measure with the new value for the probability of the basic failure event j.

Rearranging the equation slightly, yields the generalized importance measure, $I_j^G(q_j^n)$:

$$I_j^G\left(q_j^n\right) = \frac{F_j^n}{F} = \left\{ F\left[q_j = 1, \underline{q}(t)\right] - F\left[q_j = 0, \underline{q}(t)\right] \right\} \left(\frac{q_j^n - q_j}{F} \right) + 1 \quad (6.22)$$

Further rearrangement and the use of the importance measures definitions previously introduced yields:

$$I_j^G\left(q_j^n\right) = \frac{F_j^n}{F} = I_j^{cr}\left(\frac{q_j^n}{q_j} \right) + \frac{1}{RRW_j} \quad (6.23)$$

This importance measure is considered general since q_j^n can take any value and is not restricted to a value of 0 or 1 as is the case in the definitions of RRW, RAW, I^{FV}. Since the generalized importance is defined for all values of q_j^n, a continuous relationship exists between q_j^n

and the generalized importance measure. Such relationship is referred to as the 'risk impact curve'. The relationship is linear with a slope equal to the criticality importance and the y-axis intercept equal to the inverse of the risk reduction worth. Furthermore, when q_j^n equals unity the generalized importance measure attains the value of the risk achievement worth. Fig. 6.1 illustrates this relationship [9].

Fig. 6.1 Risk impact curve [9]

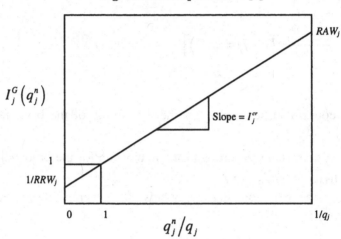

6.8 Importance measures for multiple basic events

Another important limitation of the classic importance measures previously described is that they rank only individual components or basic events whereas they are not directly applicable to combinations or groups of components or basic events. In practice different basic events may, for example, represent different modes of failure or unavailability of a single component and in order to determine the importance of such component one has to consider all the related basic events as a group. Furthermore, many risk-informed applications deal with evaluating the risk change associated to changes in the plant technical specifications (surveillance and/or test frequencies, etc.); such changes may indeed impact a group of components.

In general there is no simple relation between the importance measures of the individual components and the group taken as a whole.

Suppose that the expression of the indicator variable T of the top event of a system, in terms of the indicator variables of its basic Boolean events A, B, C_1, C_2, C_3, C_4, D, E, F, G, H is of the form [9]:

$$T = AB(C_1 + C_3) + DE(C_2 + C_4) + F(C_1 + C_3)(C_2 + C_4) + GH \qquad (6.24)$$

where as usual the product of two indicator variables denotes the intersection of the two associated events and the sum denotes the union [14]. Let us calculate some importance measures for the components of group C, constituted by two pairs of components in series, C_1-C_3 and C_2-C_4.

6.8.1 Risk achievement worth

For a single component, the calculation of the risk achievement worth entails calculating the risk of the system when the failure event of the considered component is verified with certainty. For example, the expression for RAW(C_1) in terms of basic events is [9]:

$$RAW(C_1) = \frac{AB(1+C_3)+DE(C_2+C_4)+F(1+C_3)(C_2+C_4)+GH}{AB(C_1+C_3)+DE(C_2+C_4)+F(C_1+C_3)(C_2+C_4)+GH} \quad (6.25)$$

where the indicator variable C_1 has been replaced by the value which indicates occurrence of the associated event of failure of the component C_1. Then, the numerical value of RAW(C_1) can be calculated by replacing the indicator variables with the probabilities of the corresponding basic events.

Analogously, for the components of group C, the calculation of the group RAW entails calculating the system risk when the failure event of all of the components belonging to the group is verified with certainty. An approach for calculating the group RAW could be the simple setting at the value 1 of all the indicator variables related to the basic events belonging to group C. This would result in [9]:

$$RAW(C) = \frac{2AB+2DE+4F+GH}{AB(C_1+C_3)+DE(C_2+C_4)+F(C_1+C_3)(C_2+C_4)+GH} \quad (6.26)$$

Note that this way of proceeding leads to the fact that each of the two pairs of components in series C_1-C_3 and C_2-C_4 contributes with a term 2 in the system structure function. Generalizing, the effect of this simple substitution is that for n components in series, the unavailability of the series group is n times the unavailability of the single component, which has been set to 1 and thus it leads to an unavailability equal to n. Indeed, this is not the correct approach to calculate the system risk when all of the failure events of the group are verified: the effect of this simple substitution is the generation of non-minimal cutsets.

Another approach could be to evaluate the RAW for each of the basic events and then add all the resulting RAWs. This approach, however, has problems similar to the above formulation since the numerator of this sum would contain a contribution of $4GH$, since each of the four single-component RAWs would contribute a term GH.

The correct way of proceeding for calculating the group RAW is to evaluate the structure function as a probability equation with the appropriate Boolean reduction [14]. The step-by-step procedure to be followed is:

1. Take the structure function of the top event corresponding to the measure of interest (CDF, LERF, unreliability, etc).
2. Rename the basic events in the group under investigation so that they all have the same identifier.
3. Boolean-reduce the thereby obtained structure function.
4. Calculate the risk index for the new basic event with its value set at unity.

This gives an importance measure for the group. In this case, the group is totally correlated in its effect, since the probabilities of the individual members of the group are set to one. The difference between this approach and the previous one of taking the group event probabilities individually to one in the not-reduced expression for T can be seen in the following.

The substitution of C_1, C_2, C_3, C_4 by C and the re-reducing of the equation (6.24) would give [9]:

$$T = ABC + DEC + FC + GH \qquad (6.27)$$

and

$$RAW(C) = \frac{AB + DE + F + GH}{AB(C_1 + C_3) + DE(C_2 + C_4) + F(C_1 + C_3)(C_2 + C_4) + GH} \qquad (6.28)$$

It can be easily seen that following this approach, the group RAW cannot be expressed simply in terms of combinations of the RAW measures for the individual members of the group. A consequence of this fact is that the risk equation must be re-reduced and evaluated each time a group RAW is to be evaluated.

6.8.2 Birnbaum importance measure

The Birnbaum importance measure for an individual basic event is evaluated by

$$I_j^B = F\Big[q_j = 1, \underline{q}(t)\Big] - F\Big[q_j = 0, \underline{q}(t)\Big] \qquad (6.29)$$

For the single j-th basic event, the sensitivity of the risk measure to the probability of that event, q_j, can be parameterized as

$$F\big(q_j\big) = I_j^B q_j + F\Big[q_j = 0, \underline{q}(t)\Big] \qquad (6.30)$$

The meaning of the Birnbaum importance measure therefore is that it represents the sensitivity coefficient of the risk measure to the probability of that basic event and provides one way of looking at the defense-in-depth issue in a probabilistic sense.

One may think of applying the procedures discussed above for the RAW to produce meaningful Birnbaum importance measures of groups of events. Then in the first case, for example, in evaluating a group importance using the not-reduced structure function (6.24) above, if a substitution of 1 were made for each member C_k of the group C, the resulting 'group Birnbaum measure' would be [9]:

$$I_C^B = 2AB + 2DE + 4F \qquad (6.31)$$

On the contrary, following the same substitution and re-reduction procedure, which leads to $T=ABC+DEC+FC+GH$, as suggested above for the RAW, would result in a 'group Birnbaum measure' of $AB+DE+F$. That neither of these Birnbaum importance measures is an appropriate sensitivity measure will be shortly shown in the following Section 6.9.

6.8.3 Fussell-Vesely importance

The Fussell-Vesely measure of importance for a single basic event represents the fraction of the risk measure to which the basic event contributes, i.e. it is the sum of the cutsets involving such basic event divided by the sum of all the cutsets. The Fussell-Vesely measure obtained by including all cutsets that contain one or more basic events of the group C is given by [9]:

$$I_C^{FV}(C) = \frac{AB(C_1+C_3)+DE(C_2+C_4)+F(C_1+C_3)(C_2+C_4)}{AB(C_1+C_3)+DE(C_2+C_4)+F(C_1+C_3)(C_2+C_4)+GH} \quad (6.32)$$

This is a measure that assesses the contribution of the group C in such a way that any cutset that has a contribution from any one member C_k of the group is included, $k=1,2,3,4$. Note, however, that this is not the same result that would be obtained by adding the individual Fussell-Vesely measures $I_{C_k}^{FV}$, $k=1,2,3,4$. Since this measure does not involve assessing changes, but is a simple ratio of contributors, this is an appropriate measure of group importance.

6.8.4 Risk reduction worth

The risk reduction worth importance of a single basic event is the ratio of the risk value to that calculated with the probability of such basic event set to 0. Substituting 0 for each member of the group to calculate the RRW is an appropriate way to calculate group importance since, in this case there is no problem with non-minimal cutsets.

6.9 Relationship of importance measures to system risk changes

An issue that has been noted is that the importance measures are, for the most part, not directly related to the risk changes associated with the change in the system which is considered. That this is true for those importance measures which are based on taking parameter values or basic event probabilities to their extremes should be obvious. Thus, there is concern in identifying sensitivity measures related to importance measures that can fill the role of characterizing directly the change in risk, particularly when a group of components is affected by the change. As a simple example, take a cutset equation in terms of indicator variables and treat it as an algebraic equation, replacing each of the events in the group C of interest by a common indicator variable, and without performing a Boolean reduction, differentiate the equation with respect to that variable. For the equation (6.24), replacing C_k with C, $k=1,2,3,4$, the algebraic equation would become [9]:

$$T = 2ABC + 2DEC + 4FC^2 + GH \qquad (6.33)$$

and differentiating:

$$\frac{\partial T}{\partial C} = 2AB + 2DE + 8FC \qquad (6.34)$$

This is a sensitivity parameter that is valid when the changes in the value for C are small, and the approximation

$$\Delta E[T] = \Delta F = \frac{\partial F}{\partial C} \Delta C \qquad (6.35)$$

is appropriate if the impact of the change on each member of the group is the same. This sensitivity parameter is, however, different from any of the importance measures presented in the previous sub-chapters. As the magnitude of the changes in C increases, higher order derivatives are needed to assess the change in F. Thus, it can be concluded that the sensitivity of risk to a multi-component change cannot easily be related to single-component importance measures.

6.10 The Differential Importance measure (DIM)

As highlighted in the previous Sections, a limitation of the above mentioned importance measures is that they rank only individual components or basic events whereas they are not directly applicable to combinations or groups of components or basic events [9]. To partially overcome this limitation, the Differential Importance Measure, DIM, has been introduced for use in risk-informed decision making [12]. The DIM is a first-order sensitivity measure that ranks the parameters of the risk model according to the fraction of the total change in the risk that is due to a small change in the parameters' values, taken one at a time. The DIM bears an important property of additivity: the DIM of a group of components or basic events is the sum of the DIMs of the single components or basic events of the group.

In what follows, we briefly recall the concepts underlying the definition of the Differential Importance Measure DIM introduced in [12], which the reader should consult for further details.

Consider the generic risk metric F. In general, the risk metric of interest can be expressed as a function $F(p_1, p_2, ..., p_{Np})$ of the parameters p_i, $i = 1, 2, ..., N_p$ of the underlying stochastic model (components failure rates, repair rates, ageing rates, maintenance intervals, human error probabilities, etc.). The total variation of the function of interest due to small variations in its parameters, one at a time, is given by the differential

$$dF = \frac{\partial F}{\partial p_1} \cdot dp_1 + \frac{\partial F}{\partial p_2} \cdot dp_2 + ... + \frac{\partial F}{\partial p_{N_P}} \cdot dp_{N_P} \qquad (6.36)$$

The DIM of parameter p_i is then defined as the fraction of total change in F that is due to a change in the parameter value

$$\mathrm{DIM}(p_i) = \frac{dF_{p_i}}{dF} = \frac{\dfrac{\partial F}{\partial p_i} \cdot dp_i}{\dfrac{\partial F}{\partial p_1} \cdot dp_1 + \dfrac{\partial F}{\partial p_2} \cdot dp_2 + ... + \dfrac{\partial F}{\partial p_{N_P}} \cdot dp_{N_P}} \qquad (6.37)$$

Because of its definition, once all the individual sensitivities $\partial F/\partial p_i$, $i=1,2,...,N_p$ have been computed, the DIM enjoys the additivity property, i.e. the DIM of a subset of parameters $p_i, p_j,...,p_k$, is the sum of the DIMs of the individual parameters:

$$
\text{DIM}\left(p_i, p_j, ..., p_k\right) = \frac{\dfrac{\partial F}{\partial p_i} \cdot dp_i + \dfrac{\partial F}{\partial p_j} \cdot dp_j + ... + \dfrac{\partial F}{\partial p_k} \cdot dp_k}{dF} \tag{6.38}
$$

$$
= \text{DIM}\left(p_i\right) + \text{DIM}\left(p_j\right) + ... + \text{DIM}\left(p_k\right)
$$

Viewing the definition of DIM in Eq. (6.37) in terms of a limit for the parameter variation going to zero, allows defining the operational steps for its computation. Two different hypotheses can be considered:

1) all the parameters change by the same small value (uniform changes);
2) the parameters are changed by the same percentage (uniform percentage changes).

Under hypothesis 1), $\text{DIM}(p_i)$ measures the importance of parameter p_i with respect to a small equal change in all parameters; under hypothesis 2), $\text{DIM}(p_i)$ measures the importance of parameter p_i when all the parameters are changed by the same fraction of their nominal values.

Clearly, the two assumptions address different situations and should lead to different importance values. The conditions under which to apply one hypothesis or the other depend on the problem and risk metric model at hand. In particular, when investigating the effects of changes at the parameter level, hypothesis 1) cannot be used since the parameters may have different dimensions (e.g. failure rates have inverse-time units, maintenance intervals have time units and human error probabilities are dimensionless numbers).

Example 6.5 [20]

An application of DIM is considered with reference to the Containment Spray Injection System (CSIS) of a nuclear power plant. The application shows the potential of using the DIM for identifying the relevant weaknesses of a system. The results obtained have been applied to guide the choice of a change in the system for improving its reliability/availability characteristics. An analysis of the results confirms the different physical points of view of the two hypotheses of parameter perturbation underlying the operative computation of DIM.

The function of the Containment Spray Injection System (CSIS) is to deliver cold water containing boron through spray heads from the Refuelling Water Storage Tank (RWST) to the containment volume during the first half hour after a large Loss Of Coolant Accident (LOCA). Refer to [21] for a comprehensive description of the system. The principal objective of CSIS is to reduce the pressure in the containment. The CSIS also provides the preferred path for delivery of sodium hydroxide to the containment for initial fission product removal. Fig. 6.2 shows a simplified flow diagram of the system. The CSIS consists of two redundant spray subsystems from the RWST to the containment. The valves colored in black in Fig. 6.2 are normally closed during plant operation. In order to operate both subsystems of the CSIS, valves V_5 or V_6 and V_7 or V_8 must be opened and pumps P_1 and P_2 must be started. In the event of a large LOCA this would normally be done by a signal from the Consequence Limiting Control System (CLCS). It should be noted that valves V_1 and V_3 also receive a CLCS signal to prevent those valves from being closed during the CSIS operation or to open them should they have been inadvertently closed.

Fig. 6.2 CSIS simplified flow diagram [21]

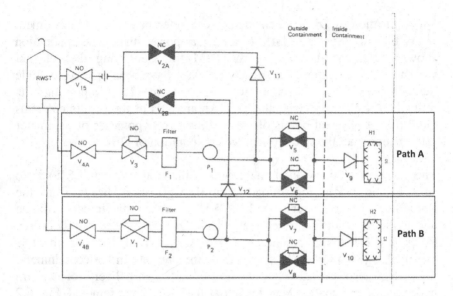

The CSIS is designed on the following basis:

    a. Either spray subsystem S_1 or S_2 will provide sufficient spray to the containment atmosphere.

    b. The CSIS is required to function only until the water supply in the RWST is exhausted.

The CSIS is considered to fail its function when it is incapable of delivering spray fluid from the RWST to the containment atmosphere at a rate at least equivalent to the full delivery from one of the two containment spray pumps. The fault tree for the CSIS failure event is reported in Fig. 6.3. The unavailability data of the basic events are reported in Table 6.6, with the original reference coding from [21]. The minimal cut sets can be readily found by inspection of the fault tree. There are three first-order cut sets, one consisting of a failure related to the RWST (Event 1 in Table 6.6), which is the only water supply for the CSIS, whereas the others are common mode failures (Events 32 and 33 in Table 6.6). The first common mode failure refers to the CLCS and accounts for the miscalibration of several sensors that prevent the proper

CLCS signal from reaching the CSIS in the event of a LOCA. The second common mode failure refers to the possibility that both CSIS flow recirculation valves V_{2A} and V_{2B} were left open after the monthly pump test due to an operator error. Several second-order cut sets also contribute to the CSIS unavailability, arising from the combination of all the failure events related to one of the two injection paths with all those of the other.

Table 6.6 Unavailability data of the failure events of the CSIS [21]

	Event/component	Code [21]	Unavailability, q_i	Occurrence rate, λ_i [d^{-1}]
1	RWST Vent Plugged	CVT0001P	$4.40 \cdot 10^{-7}$	$1.21 \cdot 10^{-9}$
	Failures on path A			
2	V_{4A} left closed	CXVA004X	$1.00 \cdot 10^{-3}$	$2.74 \cdot 10^{-6}$
3	MOV V_3 closed	CMV100AC	$1.00 \cdot 10^{-4}$	$2.74 \cdot 10^{-7}$
4	V_{2A} left opened	CXVA002X	$1.00 \cdot 10^{-2}$	$2.75 \cdot 10^{-5}$
5	Filter F_1 plugged	CFLA01AP	$1.10 \cdot 10^{-4}$	$3.01 \cdot 10^{-7}$
6	Motor Drive Clutch disengaged (Pump P_1)	CCL1A01G	$3.00 \cdot 10^{-7}$	$8.22 \cdot 10^{-7}$
7	Control circuit failure causes P_1 not to start	CST1A01F	$1.00 \cdot 10^{-3}$	$2.74 \cdot 10^{-6}$
8	P_1 fails to start	CPMA01AA	$1.00 \cdot 10^{-3}$	$2.74 \cdot 10^{-6}$
9	P_1 discontinuous running	CPMA01AF	$1.50 \cdot 10^{-5}$	$4.11 \cdot 10^{-8}$
10	Check valve V_{10} failed closed	CCVA001C	$1.00 \cdot 10^{-4}$	$2.74 \cdot 10^{-7}$
11	Spray system A nozzles plugged	CNZA001P	$1.30 \cdot 10^{-4}$	$3.56 \cdot 10^{-7}$
12	CLCS train A fails to command Pumps & Valves	GCL01	$4.60 \cdot 10^{-3}$	$1.26 \cdot 10^{-5}$
13	Insufficient power (EPS failure)	JD00	$4.10 \cdot 10^{-5}$	$1.12 \cdot 10^{-7}$
14	Insufficient power (EPS failure)	JK00	$1.10 \cdot 10^{-6}$	$3.01 \cdot 10^{-7}$
15	Unavailability due to test	No code	$1.94 \cdot 10^{-3}$	$5.32 \cdot 10^{-6}$
16	Unavailability due to maintenance	No code	$2.20 \cdot 10^{-3}$	$6.06 \cdot 10^{-6}$

Failures on path B

17	V$_{4B}$ left closed	CXVB004X	$1.00 \cdot 10^{-3}$	$2.74 \cdot 10^{-6}$
18	MOV V$_1$ closed	CMV100BC	$1.00 \cdot 10^{-4}$	$2.74 \cdot 10^{-7}$
19	V$_{2B}$ left opened	CXVB002X	$1.00 \cdot 10^{-2}$	$2.75 \cdot 10^{-5}$
20	Filter F$_2$ plugged	CFLB01AP	$1.10 \cdot 10^{-4}$	$3.01 \cdot 10^{-7}$
21	Motor Drive Clutch disengaged (Pump P$_2$)	CCL1B01G	$3.00 \cdot 10^{-4}$	$8.22 \cdot 10^{-7}$
22	Control circuit failure causes P$_2$ not to start	CST1B01F	$1.00 \cdot 10^{-3}$	$2.74 \cdot 10^{-6}$
23	P$_2$ fails to start	CPMB01BA	$1.00 \cdot 10^{-3}$	$2.74 \cdot 10^{-6}$
24	P$_2$ discontinuous running	CPMB01BF	$1.50 \cdot 10^{-5}$	$4.11 \cdot 10^{-8}$
25	Check valve V$_9$ failed closed	CCVB001C	$1.00 \cdot 10^{-4}$	$2.74 \cdot 10^{-7}$
26	Spray system B nozzles plugged	CNZb001P	$1.30 \cdot 10^{-4}$	$3.56 \cdot 10^{-7}$
27	CLCS train B fails to command Pumps & Valves	GCL02	$4.60 \cdot 10^{-3}$	$1.26 \cdot 10^{-5}$
28	Insufficient power (EPS failure)	JC00	$4.10 \cdot 10^{-5}$	$1.12 \cdot 10^{-7}$
29	Insufficient power (EPS failure)	JJ00	$1.10 \cdot 10^{-6}$	$3.01 \cdot 10^{-9}$
30	Unavailability due to test	No code	$1.94 \cdot 10^{-3}$	$5.32 \cdot 10^{-6}$
31	Unavailability due to maintenance	No code	$2.20 \cdot 10^{-3}$	$6.06 \cdot 10^{-6}$

Common Mode Failures

32	CLCS signal fail to reach CSIS	No code	1.00E-03	2.74E-06
33	Both V$_{2A}$ and V$_{2B}$ left open after test	No code	9.00E-04	2.47E-06

In Fig. 6.4 and Fig. 6.5 we report the time-dependent behaviour of the DIM computed by MC simulation (10^7 trials), with parameters changed under hypothesis H1 and H2, respectively. Then, for sake of clarity, Fig. 6.6 and Fig. 6.7 report only the values of the DIM at one year. The number of MC trials is 10^7 in both cases. The CPU time required for the simulation was of about two minutes on an ATHLON 1400 MHz processor. The additional burden in the simulation due to the computation of the 33 (number of parameters) × 36 (time points) first order sensitivities was of a factor 1.5.

Fig. 6.3 Fault tree for the CSIS, adapted from [21]

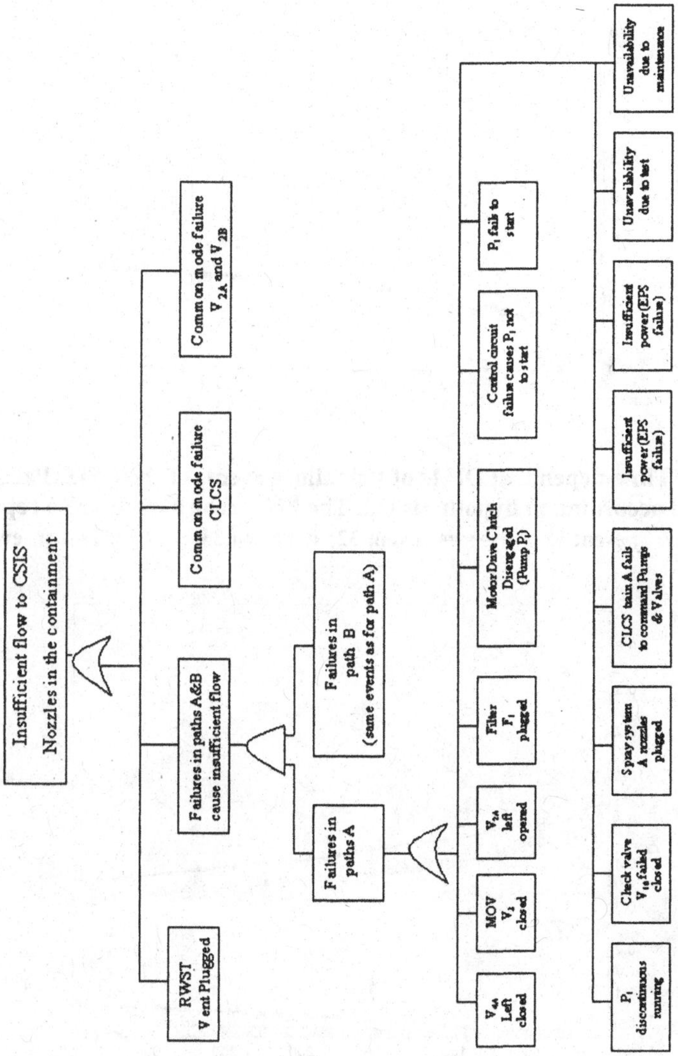

Fig. 6.4. Time-dependent DIMs of the failure events of the CSIS. Parameter changed according to hypothesis H1. The MC error bars are also reported.
(⋯⋯⋯⋯ : event 1, ▬▬▬ : event 32, o : event 33; ——: other events)

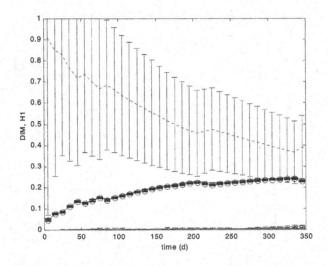

Fig. 6.5. Time-dependent DIMs of the failure events of the CSIS. Parameter changed according to hypothesis H2. The MC error bars are also reported.
(⋯⋯⋯⋯ : event 1, ▬▬▬ : event 32, o : event 33; ——: other events)

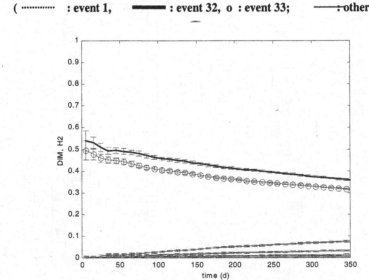

The ranking produced by the DIM at one year under hypothesis H1 (Fig. 6.6) assigns the highest importance to the three events constituting the three first-order cut sets: the RWST vent plugged and the two common mode failures of operator errors on CLCS calibration and after test of valves V_{2A} and V_{2B}. Then, the other events contributing to the second order cut sets are assigned low importance, without much difference among themselves. The reason for this ranking stands in that, under hypothesis H1, the DIM gives indications on the relevance of an event with respect to its logical role in the system, regardless of the probability that the event actually occurs. This viewpoint is similar to that of the Birnbaum IM. As a consequence, event 1 in Table 6.6 is ranked as the other two common mode failures even though its occurrence rate is four orders of magnitude lower. The very low likelihood of such event is also the reason for the great uncertainty in the estimate of its DIM.

In Fig. 6.7 we report the ranking according to the DIM under the hypothesis H2 at one year. Compared to that produced under hypothesis H1, such ranking reflects that under hypothesis H2 the logical importance of the events is weighted by their actual occurrence probability. Thus, under such hypothesis, only common mode failures are assigned high importance, while the plugging of the RWST vent drops to the group of events of lowest criticality. Furthermore, when adopting hypothesis H2, the DIM allows discriminating the relative importance among the events contributing to the second order cut sets. Indeed, events 4 and 19, i.e. valves V_{2A} and V_{2B} respectively left open, stand out from the low importance group, due to the their relatively high occurrence probability (Table 6.6). Likewise, events 12 and 27, related to failures of the CLSC to command the proper tripping of the CSIS pumps and valves, are ranked in accordance to the fact that they have the second highest occurrence probability among the events referred to each injection path.

Fig. 6.6. DIMs of the failure events of the CSIS, evaluated at one year. Parameter changed according to hypothesis H1

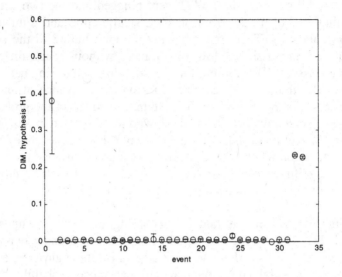

Fig. 6.7. DIMs of the failure events of the CSIS, evaluated at one year. Parameter changed according to hypothesis H2. The MC error bars are also reported.

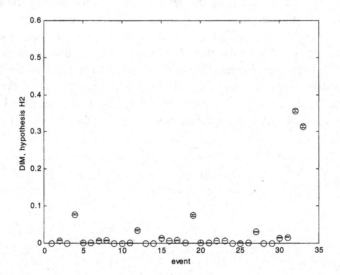

Importance measures are useful to identify possible weaknesses in the system design and management, with reference to the logic of the system or to the reliability/availability characteristics of the components and events involved. The situations identified as critical can be tackled with improvement actions, e.g. the introduction of a redundancy or of a more reliable component, aiming at reducing the criticality degree of the identified situation. The actual effectiveness of a proposed system modification must be evaluated within a risk-informed point of view to verify the net gains and losses in terms of risk and cost. This aspect will however not be considered here.

With reference to the results of the DIM for the CSIS, one can suggest the introduction of corrective actions to limit the criticality of the high-importance events. In order to prioritise the corrective actions we refer to the ranking produced under the hypothesis H2, which seems more informative for our purposes. Both common mode events are related to human errors. We analyse the effect on the system of two possible corrective actions. The first one refers to the case that the CLSC miscalibration can be detected immediately upon occurrence and a mitigating repair action is undertaken. We take, for simplicity, a constant repair rate of arbitrary value $\mu = 0.5$ d^{-1}. The second one considers the possibility of resorting to two independent testing teams for valves V_{2A} and V_{2B} so that the contribution to the failure of V_{2A} and V_{2B} due to a common mode event vanishes and only the two independent failure events 4 and 19 remain. Fig. 6.8 shows the effects on the values of the DIMs after the mitigation action has been done: as expected, the importance of the event of CLSC miscalibration is significantly reduced and the independent failures of valves V_{2A} and V_{2B} assume top relevance. The modification results in a decrease of a factor of 5 in the system unavailability.

Fig. 6.8. DIMs of the failure events of the CSIS after the mitigating actions, evaluated at one year. Parameter changed according to hypothesis H2. The MC error bars are also reported.

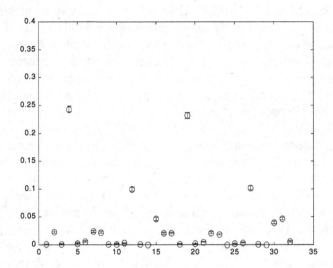

6.11 Importance measures for multi-state systems

6.11.1 Introduction

As above mentioned, classical importance measures, such as the Birnbaum, Fussell-Vesely, risk achievement worth and risk reduction worth, have been mainly applied to systems made up of binary components (i.e., components that can be in two states: functioning or faulty).

Since many components and systems are multi-state, efforts have been made to extend the evaluation of the importance of components to multi-state systems. Early progress towards the extension of the Birnbaum measure to the case of multi-state systems can be found in [16], for the case of finitely many states and in [22], for the case of continuum structure functions. Later, in [23] and [24] the Birnbaum measure has been applied to the case of multi-state systems composed by binary components and to components with dual failure-modes, respectively.

Importance measures related to the occupancy of a given state by a component have been proposed in [16] and [25]: these measures characterize the importance of a given component being in a certain state or degrading to the neighboring state with respect to the expected system performance. The IM of a given component is, therefore, represented by a vector of values, one for each state of the component. Such representation may be of difficult interpretation to the practical reliability analyst.

Other measures have been defined in order to prioritize multi-state components with respect to the MSS availability. In [26], two measures are proposed to identify the components with the highest potential of improvement in the system availability and the components responsible for the unutilized capacity of the system. Furthermore, in [27] measures are proposed in order to characterize how much a performance level of a component is responsible for the achievement or non-achievement of a given system performance.

A generalization of some commonly used importance measures has been proposed for application to multi-state systems constituted by multi-state components [27]. Physically, these measures characterize the importance for a multi-state component of achieving a given level of performance and their definitions entail evaluating the system output performance measure when the functioning of the component of interest is restricted in performance. In [27], an analysis of the generalized measures is presented when the performance of the components is restricted according to different models and when different system output performance measures are considered.

The following sub-Section 6.11.2 formally characterizes multi-state systems. In sub-Section 6.11.3 we provide a concise literature review of the IMs proposed for MSS, addressing the different views they take on the role of a multi-state component in a MSS. Then, in sub-Section 6.11.4 the generalization of binary IMs to MSS-IMs is presented. A comparison of different IMs for MSS, from both the analytical and physical viewpoints, is offered in sub-Section 6.11.5.

6.11.2 The model of a multi-state system

Consider a system made up of n components. Let $X_j(t)$ be a random variable representing the performance level of component j at time t, $j=1,2,\ldots,n$. $X_j(t)$ can assume one of m_j+1 values:

$$x_{j0}, x_{j1}, \ldots, x_{jm_j} \quad \left(0 = x_{j0} \leq x_{j1} \leq \ldots \leq x_{jm_j}\right) \qquad (6.39)$$

The value x_{jk} is the level of performance of component j when in state k, ranging from complete failure ($x_{j0} = 0$) to the maximum performance value (x_{jm_j}).

Let $W(t)$ be a non-negative random variable representing the performance level of the system at time t. The system performance $W(t)$ is determined on the basis of the individual components' performances, $X_j(t), j=1,2,\ldots,n$ and depends on the system logic of operation of the considered system. $W(t)$ can assume one of $m+1$ values, ranging from complete system failure (state $i=0$) to perfect functioning (state $i=m$):

$$w_0, w_1, ..., w_m \quad \left(0 = w_0 \leq w_1 \leq ... \leq w_m\right) \tag{6.40}$$

In practice, MSS may be requested to work at different performance levels at different times. For example, the production of electrical and thermal power plants varies according to the daily and seasonal load demands. Assume that at time t a given level of system performance $D(t)$ is required.

The behavior of a MSS is generally judged in terms of a measure of output system performance O [15]. For example, the system steady-state availability or the system steady-state performance are frequently used. A detailed description of the various measures of MSS output performance can be found in [15]. In the following illustration of the MSS importance measures, we shall often refer to the availability at time t, which for a MSS is the probability that at that time the system has performance $W(t) \geq D$.

6.11.3 Importance measures for MSS

One of the first notions of IMs for multi-state components in MSS has been introduced in the early eighties by Griffith [16]. Consider a system made up of n components having $m+1$ possible levels of performance w_i such that $0 \leq w_0 \leq w_1 \leq ... \leq w_m$. Each component j, $j=1,2,...,n$, has m_j possible states. The performance of component j when in state k is x_{jk} ($0 \leq x_{j0} \leq x_{j1} \leq ... \leq x_{jm_j}$).

Physically, the measure of the Griffith's importance of component j being in state k, $I_k^G(j)$, represents the variation in the expected system performance due to a degradation of component j from the performance state k to the performance state $k-1$. Note that, when applied to binary systems, the Griffith's IM reduces to the Birnbaum's importance measure.

In this sense, the Griffith's importance measure allows one to identify those performance states of the components for which a single-step decrement in performance has major effects on the system.

The Griffith's importance measure of component j is the vector $\underline{I}^G(j) = \left(I_1^G(j), \ldots, I_{m_j}^G(j) \right)$. In [16], it is shown that $\underline{I}^G(j)$ can be interpreted as the rate of improvement of the MSS performance following an improvement in the performance of its multi-state components.

Later, the performance utility importance function $I_k^U(j)$ has been introduced in order to identify which levels of components performance contribute the most to the system performance [25]. Such information is indeed not easily retrievable from the Griffith's IM.

The performance utility importance function, $I_k^U(j)$, of state k of component j is the expected value of the system performance when component j resides in state k times the probability that the component j actually resides in that state k. In this sense, $I_k^U(j)$ can be interpreted as the contribution of state k to the overall system performance.

The utility importance of component j is defined by the vector $I_k^U(j)$ of the importances of its individual states, i.e. $\underline{I}^U(j) = \left(I_0^U(j), I_1^U(j), \ldots, I_{m_j}^U(j) \right)$. The performance utility importance function is useful to determine which state of a component contributes the most to the overall system performance, compared to the other states of that component. If a state k of component j has a high value of $I_k^U(j)$, it significantly contributes to the system performance. Note that, by definition, a state k of component j is important according to the measure I^U either if the system has high performances when component j is in state k (high value of $E[W_{jk}]$) or if the probability p_{jk} of component j being in state k is high.

Other importance measures characterize the role of multi-state components with respect to the MSS availability. In particular, in [26] the system availability improvement potential, I_1 and expected unutilized capacity of component j, I_2 are introduced. The system availability improvement potential of component j, $I_1(j)$, indicates which components should receive attention in order to increase the system availability the most. This measure is useful in the identification of system bottlenecks. Physically, $I_1(j)$ equals the variation in the system availability obtained by fixing the performance of component j to its highest achievable one, e.g. 100%. It can be verified that $I_1(\ j)$ equals the probability that component j acts as a system bottleneck.

The expected unutilized capacity of component j, $I_2(\ j)$, expresses how much the performance of the component can be reduced without effects on the system availability. This measure is useful in the system design phase to identify components having too much or too little extra performance with respect to that actually required by the system.

In [27] two other importance measures are introduced, characterizing the contribution to the system availability of state k of component j. The first measure, $I^{M1}(j, k, t)$, is the probability that the performance $W(t)$ of the system at time t is smaller than the required performance $D(t)$ when component j is in its lower performance state 0 and greater than $D(t)$ just when component j is in state k. Such measure characterizes how much state k of component j is responsible for the system providing at least the required performance level $D(t)$. The second measure, $I^{M2}(j, k, t)$, is the probability that the performance $W(t)$ of the system at time t is smaller than the required performance $D(t)$ when component j is in the state k-1 with performance just lower than that of state k and that $W(t) \geq D(t)$ when component j is in its best performing state $k=m_j$. Such measure characterizes how much state k of component j is responsible for the missed delivery by the system of the required performance $D(t)$.

6.11.4 Importance measures based on limitations on the performance of multi-state components

The definitions of the Birnbaum, the Fussell-Vesely, the risk achievement worth and the risk reduction worth IMs have been generalized to multi-state systems [28]. The question that these generalized IMs address concerns the importance, with respect to the considered MSS output performance measure W, that a given component j achieves a pre-defined level of performance α. This would give the analyst an indication of the components' performance levels that are most critical for the system performance. To this aim, Multi-State System Importance Measures have been introduced.

We denote by $k_{j\alpha}$ the state in the ordered set of states of component j whose performance $x_{jk_{j\alpha}}$ is equal to or immediately below α, i.e. $x_{jk_{j\alpha}} \leq \alpha < x_{jk_{j\alpha}+1}$. Then, we introduce:

- $W_j^{\leq\alpha} = W\left(\underline{X} \big| X_j \leq \alpha \text{ in } [0,\tau]\right)$: system output performance when the performance X_j of the j-th component is restricted to be below or equal to α (i.e., component j is restricted in states $k \leq k_{j\alpha}$) in $t \in [0,\tau]$.

- $W_j^{>\alpha} = W\left(\underline{X} \big| X_j > \alpha \text{ in } [0,\tau]\right)$: system output performance when the performance X_j of the j-th component is restricted to be above α (i.e., component j is restricted in states $k > k_{j\alpha}$) in $[0,\tau]$.

By so doing, the complete ordered set of states of the generic j-th component is divided into two ordered subsets, thus re-introducing a collectively binary logic for the states *functioning* above performance level α and *faulty* below level α, respectively. In this framework, the following generalized IMs can be defined as follows:

Birnbaum measure of α-level

$$bW_j^\alpha = W_j^{>\alpha} - W_j^{\leq\alpha} \tag{6.41}$$

The bW_j^α is the maximum change in system output performance W when the performance of component j is changed from always above the α-level ($X_j>\alpha$, i.e., states $k > k_{j\alpha}$) to always below or equal to the α-level of performance ($X_j \leq \alpha$, i.e., states $k \leq k_{j\alpha}$).

Fussell-Vesely measure of α-level

$$fW_j^\alpha = \frac{W - W_j^{\leq\alpha}}{W} \tag{6.42}$$

The fW_j^α is the ratio of the decrement in the system output performance W due to the component j operating with a level of performance below or equal to α ($X_j \leq \alpha$, i.e., states $k_j \leq k_{j\alpha}$) in $[0, \tau]$ to the nominal value of W.

Achievement Worth of α-level

$$aW_j^\alpha = \frac{W_j^{>\alpha}}{W} \tag{6.43}$$

The aW_j^α depends on the system output performance achieved by the system when component j is obliged to operate with a performance above α ($X_j > \alpha$, i.e., states $k > k_{j\alpha}$) in $[0, \tau]$.

Reduction worth of α-level

$$rW_j^\alpha = \frac{W}{W_j^{\leq \alpha}} \tag{6.44}$$

The rW_j^α represents the reduction in W which can be achieved when the output performance of component j is maintained below or equal to level α ($X_j \leq \alpha$, i.e., states $k \leq k_{j\alpha}$). Also in the case of MSS, rW^α and fW^α produce the same ranking of component importance.

The above definitions hold also for continuous-states components and systems. In the case of continuous states, the α-level can continuously assume any intermediate value within its range (e.g. $\alpha \in [0\%, 100\%]$) and the state indicator variable k can vary continuously within $[0, m_j]$, $j=1,2,\ldots,n$.

Example 6.6 [26]

Let us consider a system made up of a series of $\eta = 2$ macro-components (nodes), each one performing a given function (Fig. 6.9). Node 1 is constituted by $n_1 = 2$ components in parallel logic, whereas node 2 is constituted by a single component ($n_2 = 1$) so that the overall number of components in the system is $n = \sum_{l=1}^{\eta} n_l = 3$. The mission time τ is 1000 hours.

Fig. 6.9 Reliability block diagram of the system considered

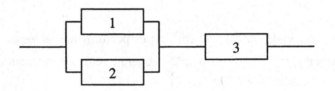

For each component $i=1,2,3$ there are $m_i = 5$ possible states, each one corresponding to a different hypothetical level of performance $x_{i,j}$, $j=1,2,\ldots,5$. Each component is assumed to move stochastically from one state j to another state k, according to exponential time distributions with rate $\lambda_{i,j \rightarrow k}$ (h^{-1}). For each component $i=1,2,3$, we then have a transition matrix Λ^i of the values of the transition rates:

$$\Lambda^1 = \begin{bmatrix} - & 5 \cdot 10^{-3} & 0 & 0 & 5 \cdot 10^{-4} \\ 5 \cdot 10^{-3} & - & 5 \cdot 10^{-3} & 0 & 6 \cdot 10^{-3} \\ 0 & 5 \cdot 10^{-3} & - & 5 \cdot 10^{-3} & 8 \cdot 10^{-3} \\ 0 & 0 & 5 \cdot 10^{-3} & - & 8 \cdot 10^{-3} \\ 1 \cdot 10^{-2} & 5 \cdot 10^{-3} & 5 \cdot 10^{-3} & 5 \cdot 10^{-3} & - \end{bmatrix} \qquad (6.45)$$

$$\Lambda^2 = \begin{bmatrix} - & 5 \cdot 10^{-3} & 0 & 0 & 1.5 \cdot 10^{-3} \\ 5 \cdot 10^{-3} & - & 5 \cdot 10^{-3} & 0 & 2 \cdot 10^{-3} \\ 0 & 5 \cdot 10^{-3} & - & 5 \cdot 10^{-3} & 3 \cdot 10^{-3} \\ 0 & 0 & 5 \cdot 10^{-3} & - & 4 \cdot 10^{-2} \\ 1 \cdot 10^{-2} & 5 \cdot 10^{-3} & 5 \cdot 10^{-3} & 5 \cdot 10^{-3} & - \end{bmatrix} \qquad (6.46)$$

$$\Lambda^3 = \begin{bmatrix} - & 5\cdot10^{-4} & 0 & 0 & 5\cdot10^{-5} \\ 5\cdot10^{-3} & - & 5\cdot10^{-4} & 0 & 6\cdot10^{-5} \\ 0 & 5\cdot10^{-3} & - & 5\cdot10^{-4} & 7\cdot10^{-5} \\ 0 & 0 & 5\cdot10^{-3} & - & 8\cdot10^{-5} \\ 1\cdot10^{-1} & 5\cdot10^{-2} & 5\cdot10^{-2} & 5\cdot10^{-2} & - \end{bmatrix} \qquad (6.47)$$

Table 6.7 gives the values of the performances $x_{i,j}$ (in arbitrary unit) of the three components in correspondence of all the possible states $j=1,2,\ldots,5$. Note that state 5 corresponds to zero-performance, i.e. component failure.

Table 6.7 Components' performance data

Component (i)	Performance ($x_{i,j}$)				
	$j=1$	$j=2$	$j=3$	$j=4$	$j=5$
1	80	60	40	20	0
2	80	60	40	20	0
3	100	75	50	25	0

The output performance $W_{\underline{j}}$ associated to the system state $\underline{j}=(j_1, j_2, \ldots, j_n)$ is obtained on the basis of the performances $x_{i,j}$ of the components $i=1,2,\ldots,n$ constituting the system. More precisely, we assume as in [23] that the performance of each node l constituted by n_l elements in parallel logic, is the sum of the individual performances of the components and that the performance of the two-nodes series system is that of the node with the lowest performance, which constitutes the 'bottleneck' of the system. For example, with reference to the system configuration $\underline{j}^* = (1, 3, 2)$, the first node is characterized by a value of the performance equal to $x_{1,1} + x_{2,3} = 120$, while the second node has performance $x_{3,2} = 75$. This latter node determines the value of the system performance $W_{\underline{j}^*} = 75$.

Let Γ_i^α be the set of states of component i characterized by a performance level below or equal to α and $\overline{\Gamma_i^\alpha}$ be the set of states of component i characterized by a performance level above α (complement set of Γ_i^α). Then, $\overline{W_i}^{\leq\alpha} = \overline{W}(j_i \in \Gamma_i^\alpha \text{ in } [0,\tau])$ is the system mean performance over the period τ when the performance of the i-th component is below or equal to α (i.e., $j_i \in \Gamma_i^\alpha$) in $[0,\tau]$ and $\overline{W_i}^{>\alpha} = \overline{W}(j_i \in \overline{\Gamma_i}^\alpha \text{ in } [0,\tau])$ is the system mean performance over τ when the performance of the i-th component is above α (i.e., $j_i \in \overline{\Gamma_i^\alpha}$) in $[0,\tau]$.

The system mean performances $\overline{W}$, $\overline{W_i}^{\leq\alpha}$, $\overline{W_i}^{>\alpha}$, $i=1,2,\ldots,n$, of interest for the evaluation of the MSS importance measures have been estimated by the Monte Carlo simulation of the system stochastic evolution in time (chapter 2).

Fig. 6.10 reports the results pertaining to the calculation of the MSS performance importance measures for each of the three components and all possible α-levels. Component 3 is the most important according to the measures rW_j^α, fW_j^α and bW_j^α and the least important according to aW_j^α at any α-level. The measures rW_3^α, fW_3^α are indicators of the reduction in mean system performance $\overline{W_3}^{\leq\alpha}$ due to component 3 providing at most a level of performance α. Indeed, the performance $x_{3,j}$ of component 3 in its state $j=1,2,\ldots,5$, determines the system performance which cannot exceed the value $\overline{W_3}^{\leq\alpha} = \alpha = x_{3,j}$. The large values of $bW_3^\alpha = \overline{W_3}^{>\alpha} - \overline{W_3}^{\leq\alpha}$ also follows from the above considerations. Thus, as expected, to achieve a satisfactory system performance, it is very important to assure a sufficient α-level for component 3. Note that rW_3^0 equals infinity, given that the performance reduction $\overline{W_3}^{\leq 0}$ of component 3 at 0-level (i.e., with component 3 always in the zero performance state $j=5$, $x_{3,5}=0$), equals zero (correspondingly, $fW_3^0 = 1$). On the contrary, the small values of the performance achievement worth measure, obtained at any α-level indicate that actions for improving the performance of components above α are more effective if devoted to component 1 or 2. Indeed, component 3 is already characterized by high average performance over the mission time $\tau(\bar{x}_3 = 97.531)$.

Fig. 6.10 MSS performance importance measures as a function of the reference α-level (state j) for each component

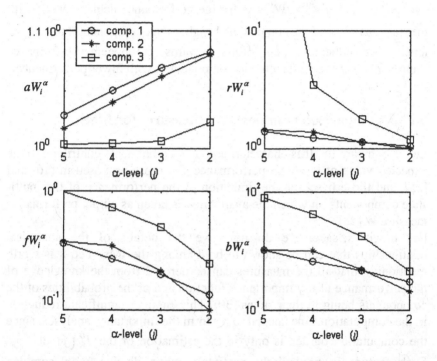

As for the relative ranking of components 1 and 2, at lower component performance levels, the performance reduction measures rW^α and fW^α and the Birnbaum measure bW^α indicate that component 2 is more important than component 1. This is due to the fact that the average performance over the mission time τ of component 2 is higher ($\bar{x}_2 = 59.085$) than that of component 1 ($\bar{x}_1 = 58.116$) and when the performance of one of the two components in the parallel logic node is forced to the 0-level, the performance of the node is entirely determined by the other component so that $\overline{W}_1^{\leq 0} > \overline{W}_2^{\leq 0}$ because $\overline{x}_2 > \overline{x}_1$ and, correspondingly, $rW_1^0 < rW_2^0$ and $fW_1^0 < fW_2^0$. On the other hand, at higher α-levels, the least performing component 1 more and more affects the system performance $\overline{W}_1^{\leq \alpha}$, which, at the highest α-level corresponding to $j = 1$ for $i=1,2,3$ becomes lower than $\overline{W}_2^{\leq \alpha}$ so that $rW_1^0 > rW_2^0$ and $fW_1^0 > fW_2^0$. As for the performance achievement worth measure, aW^α, it ranks component 1 higher than component 2 for any α-level: this indicates that efforts towards component performance improvement are to be devoted to component 1, with lower performance.

6.11.5 Comparison of importance measures for MSS

Let us compare the IMs characterizing how components contribute to the expected value of the MSS performance, i.e. those proposed in [16] and [25], and those based on the limitation of the performance of the multi-state components, in which the availability is taken as output performance measure W [28].

For brevity's sake, we do not give the details of the analytical relationships that can be shown to hold among the IMs. Yet, it is worth mentioning that all the measures can be derived from the knowledge of the performance utility importance function and of the probabilities of the components being in their states [30]. This entails a significant reduction in the computation time needed to perform the importance analyses, since the computation burden is only in the estimation of the $I_k^U(j)$ and p_{jk}.

Such feature is particularly profitable when dealing with complex systems whose modeling often requires the use of time-consuming simulation codes. In these cases, there is no need to repeat the system simulation each time a different IM is considered, since the value of the IM can be derived from $I_k^U(j)$ and p_{jk}.

From the physical point of view, the IMs considered are related to the occupancy of a given state by a component. However, the various IMs refer under different perspectives to the event that a component occupies a given state or a subset of states. In particular, a distinction can be made between measures referring to 'occurring events' and to 'existing conditions' as pointed out in [29] with reference to IMs for binary states. An example of an occurring event is the failure of a component, whereas an example of an existing condition is a component being faulty. The importance of an occurring event depends both on the effect of the event on the system as well as on the probability of that event actually occurring. On the other hand, the importance of an existing condition is related to the fallbacks of that condition on the system, regardless of its occurrence probability. The $I_k^U(j)$ measure considers the event of component j residing in state k as an 'occurring event'. Indeed, by definition $I_k^U(j)$ is the expected system performance $E[W_{jk}]$ when component j resides in state k times the probability p_{jk} of component j of actually residing in that state. Instead, the other measures refer to existing conditions: the one-step degradation addressed by I^G or the confinement of the components into states with performance always below or above α are given conditions and their effect on the system are analyzed regardless of their actual occurrence probability.

IMs have different applications for decision making depending on whether they refer to occurring events or existing conditions. Consider a case in which the goal is to prioritize actions for system improvement, such as increasing inspection/maintenance frequencies or allocating redundancies. Then, it seems reasonable to consider IMs adopting the existing condition perspective: indeed, the analyst has to judge the performance of the system after the improvement has been done. For example, if the aim is that of achieving the maximum improvement in system performance, the analyst's decision should be driven by the measures aW_j^{α}; on the contrary, if the aim is that of reducing the likelihood of low- or zero-performance system configurations then one should follow the prioritization suggested by the performance reduction measures rW_j^{α}, fW_j^{α}.

On the contrary, from the perspective of the occurring event, the $I_k^U(j)$ measures identify which components in which state contribute or not to the performance of the system without any change. In this view, the system performance can be increased by acting on component j with respect to either $E[W_{jk}]$ or p_{jk}, depending on which factor is responsible for the low value of $I_k^U(j)$.

A more detailed discussion of the analytical and physical relationships holding among the measures is contained in [30].

6.12 References

[1]   Birnbaum L.W., *On the Importance of Different Elements in a Multi-Element System,* Multivariate analysis 2, New York, Academic Press, 1969.

[2]   Fussell J.B., *How to Calculate System Reliability and Safety Characteristics,* IEEE Trans. on Reliab., 1975; R-24(3); 169-174.

[3]   Meng, F.C., *Comparing the Importance of System Elements by Some Structural Characteristics,* IEEE Trans on Reliab, 1996 vol 45 no 1 pp 59-65.

[4]   Meng, F.C., *Some Further Results on Ranking the Importance of System Elements,* Reliab. Eng. and Sys. Safety, 1995; 47; 97-101.

[5]   Elsayed, E. A., *Reliability Engineering,* Addison Wesley Longman, 1996.

[6]   Høyland, A. and Rausand, M., *System Reliability Theory: Models and Statistical Methods,* John Wiley & Sons, 1994.

[7]   Hong, J. S. and Lie, C. H., *Joint Reliability-Importance of Two Edges in an Undirected Network,* IEEE Trans on Reliab 1993 vol 42 no 1 pp 17-23.

[8]   Armstrong, M. J., *Joint Reliability-Importance of Elements,* IEEE Trans on Reliab 1995 vol 44 no 3 pp 408-412.

[9]   Cheok, M. C., Parry G. W., Sherry R. R., *Use of Importance Measures in Risk Informed Applications,* Reliab. Eng. and Sys. Safety 1998; 60; 213-226.

[10] Vasseur D., Llory M., *International Survey on PSA Figures of Merit,* Reliab. Eng. and Sys. Safety, 1999; 66; 261-274.

[11] van der Borst, M, Shoonakker, H, *An Overview of PSA Importance Measures,* Reliab. Eng. and Sys. Safety, 2001; 72(3); 241-245.

[12] Borgonovo, E., Apostolakis, G. E., *A New Importance Measure for Risk-Informed Decision Making,* Reliab. Eng. and Sys. Safety, 2001; 72; 193-212.

[13] Youngblood, R.W., *Risk Significance and Safety Significance,* Reliab. Eng. and Sys. Safety, 2001; 73; pp. 121-136.

[14] Zio, E., *An Introduction to the Basics of Reliability and Risk Analysis,* Series in Quality, Reliability and Engineering Statistics, Vol.13, World Scientific, Singapore, 2007.

[15] Aven, T. *On Performance Measures for Multistate Monotone Systems,* Reliab. Eng. and Sys. Safety, 1993; 41; 259-266.

[16] Griffith, W. S. *Multistate Reliability Models*, Journal of Applied Probability, Vol. 17 (1980), pp. 735-744.

[17] Zio, E. and Podofillini, L., *A Second-order Differential Importance Measure for Reliability and Risk Applications*, SAMO 2004, Sensitivity Analysis of Model Output, March 8-11, 2004, Santa Fe, New Mexico, USA, available on CD-rom.

[18] Modarres, M., *Risk Analysis in Engineering: Probabilistic Techniques*, Tools and Trends, CRC Press, 2006.

[19] Baraldi, P., Zio, E., Compare, M., *Importance Measures in Presence of Uncertainties*, in proceedings of SSARS 2008, Gdańsk/Sopot, Poland, 2008.

[20] Marseguerra, M., Zio, E., Podofillini, L., *First-Order Sensitivity Analysis of a Nuclear Safety System by Monte Carlo Simulation*, Reliability Engineering and System Safety, 90, 2005, pp. 162-168.

[21] Wash-1400 (NUREG 75/014), *Reactor Safety Study; An Assessment of Accident Risks in U.S. Commercial Nuclear Power Plants*, Appendix 2: Fault Trees, 1975.

[22] Kim, C. and Baxter, L. A., *Reliability Importance for Continuum Structure Functions*, Journal of Applied Probability, Vol. 24 (1987), pp. 779-785.

[23] Levitin, G., Lisnianski, A., *Importance and Sensitivity Analysis of Multi-State Systems Using the Universal Generating Function Method*, Reliab. Eng. and Sys. Safety, 1999; 65; 271-282.

[24] Armstrong, M. J., *Reliability-Importance and Dual Failure-Mode Elements*, IEEE Trans. on Reliab., 1997; 46(2); 212-221.

[25] Wu, S., Chan, L., *Performance Utility-Analysis of Multi-State Systems*, IEEE Trans. on Reliab., 2003; 52(1); 14-20.

[26] Aven, T., Østebø, R., *Two New Importance Measures for a Flow Network System*, Reliability Engineering, 14, 1986, pp. 75-80.

[27] Meng, F. C., *Element-Relevancy and Characterization Results in Multi-State Systems*, IEEE Trans. on Reliab., 1993; 42 (3); 478-483.

[28] Zio, E., Podofillini, L., *Importance Measures of Multi-State Components in Multi-State Systems*, International Journal of Reliability, Quality and Safety Engineering, Vol 10, No 3 (2003) 289-310.

[29] Vesely, W. E., *Supplemental Viewpoints on the Use of Importance Measures in Risk-Informed Regulatory Applications*, Reliab. Eng. and Sys. Safety, 60 (1998) 257-259.

[30] Zio, E., Marella, M., Podofillini, L. *A Comparison of Different Importance Measures for Multistate Systems,* MMR 2004, Mathematical Methods in Reliability, June 21-25, Santa Fe, New Mexico, USA.

7. BASIC CONCEPTS OF UNCERTAINTY AND SENSITIVITY ANALYSIS

7.1 Introduction

The quantitative description of the phenomena occurring in many engineering applications requires the adoption of mathematical models which are then turned into operative computer codes for simulation. In this sense, a model provides a representation of a real system dependent on a number of hypotheses and parameters. The model can be deterministic (e.g. Newton's dynamic laws or Darcy's law for groundwater flow) or stochastic (e.g. the Poisson's model for describing the occurrence of earthquake events).

Often in practice, the system under analysis can not be characterized exactly and the knowledge of the undergoing phenomena is incomplete. This leads to uncertainty on both the values of the model parameters and on the hypotheses underlying the model structure. Such uncertainty propagates within the model and causes variability in its outputs: for example, when many values are plausible for a model parameter, the model outputs associated to the different values of the uncertain parameter will be different; the quantification and characterization of the resulting output uncertainty is of paramount importance.

Uncertainty analysis aims at propagating to the output the uncertainties of the model input parameters and structure. In the following, we shall discuss separately the uncertainty related to the imprecise knowledge of the model parameters and that related to the uncertainty of the model structure, i.e. due to the existence of alternative plausible hypotheses on the phenomena involved. While the first source of uncertainty has been widely investigated and more or less sophisticated methods have been developed to deal with it, research is still ongoing to obtain effective and agreed methods to handle the uncertainty related to the model structure.

Sensitivity analysis aims, instead, at identifying the contribution to the output uncertainty of the various model parameters and hypotheses. Opportunely defined importance indexes are typically introduced to measure the amount of output uncertainty due to the various input parameter. In several areas of applications, this analysis can be of great aid in detecting where to act on the model for effectively reducing the output uncertainty.

7.2 Local and global uncertainty analysis

Let us consider the following generic form of a model (Fig. 7.1)

$$y = m(\underline{x})$$

where $\underline{x}$ is the vector of the n variables or uncertain parameters, m represents the model structure and $\underline{y}$ is the output vector of the model m.

Furthermore, let $f_X(\underline{x})$ be the probability density function (pdf) of the uncertain vector $\underline{x}$, $p(m_i)$ the probability distribution of the alternative plausible hypotheses of the model structure, $i=1,2,...,M$, $f_Y(\underline{y})$ the pdf which describes the resulting uncertainty on the output y (Fig. 7.1). For ease of explanation, we shall consider a one-dimensional output y.

The variability of the model output y can be studied following a *local* approach of investigation of the effects on the output of variations in the input parameters values [1,2]. Typically, the approach focuses on the nominal best estimate values $\underline{x}^0$ of the model parameters and observes what happens to the output y for small variations around those values. In the limit view, the sensitivity measure of the contribution of the generic input parameter x_i to the uncertainty of the output y is the partial derivative of y with respect to x_i calculated around the nominal values of the parameters, $\underline{x}^0$:

$$\left(\frac{\partial y}{\partial x_i} \right)_{\underline{x}^0}$$

Such measure identifies the critical parameters as those whose variation leads to the most variation in the output y. The practical approach for such identification consists in perturbing one single parameter x_i at a time with small variations around its nominal value, while maintaining the others set at their respective nominal values.

The analysis is intrinsically *local* and the resulting indication can be considered valid for the characterization of the model response around the nominal values $\underline{x}^0$. The possibility of extending the results of the analysis to draw *global* considerations on the model response over the whole input variability space depends on the model *m* itself: if the model is linear or mildly non-linear, then the extension may be possible; if the model is strongly non-linear and characterized by sharp variations, the analysis is valid only locally.

Typical local approach techniques are those based on Taylor's differential analysis (e.g., *the method of moments,* subchapter 7.3) and on the *one-at-a-time* simulation in which the parameters are varied one at a time while the others remain set at their nominal values [2,3].

Fig. 7.1 The model and its sources of uncertainty

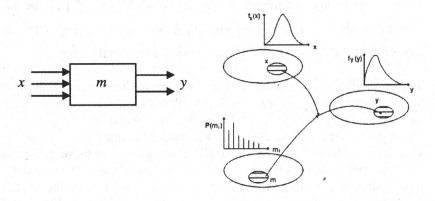

In most practical situations, models are non-linear and non-monotone: in this case, the results provided by a *local* analysis may have a limited interest. For this reason, a *global* approach to uncertainty and sensitivity analysis has been developed [2]. In this approach, the focus is directly on the output y and its uncertainty distribution $f_Y(y)$, which includes all the information about the variability of the model response, with no reference to any particular value of the input parameters (as the nominal values $\underline{x}^0$ in the local analysis).

The two principal aspects characterizing the global approach are somewhat opposite to those characteristic of the local analysis: 1) the account given to the whole variability range of the input parameters (and not only to small perturbations around the nominal values); 2) the focus on the effects resulting from varying one input parameter at a time, but while considering (on average) also the variation of the other uncertain parameters (instead of keeping them fixed to their nominal values).

In simple words, the global approach aims at allocating, on average, the variability in the output y to the various inputs and subsets of them. Conceptually, if keeping fixed the values of the input parameters in a given subset X_s, the variability of y (represented by the conditioned pdf $f_{Y|X_s}$) is considerably reduced (on average for different fixed values of X_s), such subset of input parameters is to be considered "important", i.e. the uncertainty on y is strongly influenced by the uncertainty on X_s. Since the approach does not require any particular hypothesis on the model structure, it is valid also in presence of strong non-linearities.

Many global analysis methods have been developed, such as the variance decomposition [1,4,5] (subchapter 7.7), the event and probability trees [6] (subchapter 7.4.2), the Sobol indexes [7,8] and the Fourier Amplitude Sensitivity Test (FAST) [9,10] (subchapter 7.8). The high capabilities of these methods are paid by a very high computational cost.

7.3 Approximated analytical methods: the method of moments

In practice, exact analytical methods for uncertainty propagation are not applicable, except for few "lucky" cases concerning very simple models (e.g., the combination of linear Gaussian variables or the product of lognormal variables) [11].

Nevertheless, there are approximated analytical techniques, based on the Taylor's series expansion of the model $m(\underline{x})$ around the input nominal values $\underline{x}^0$, which allow relatively straightforward practical applications. A widely common technique is the *method of moments*, which derives its name by the fact that the propagation and analysis of the uncertainties are based on the mean, variance and, possibly, on higher order moments, of the probability distributions [3].

The analysis starts with the Taylor series expansion of the model function $y=m(\underline{x})$ around the mean values of the input variables, taken as the nominal values $\underline{x}^0$:

$$y = m(\underline{x}^0) + \sum_{i=1}^{n}(x_i - x_i^0)\left(\frac{\partial y}{\partial x_i}\right)_{\underline{x}^0} + \frac{1}{2}\sum_{i=1}^{n}\sum_{j=1}^{n}(x_i - x_i^0)(x_j - x_j^0)\left(\frac{\partial^2 y}{\partial x_i \partial x_j}\right)_{\underline{x}^0} + ... \quad (7.3)$$

If the function in the interested range is reasonably "smooth", the higher order terms can be neglected. By computing the expected value $E[y]$ from (7.3) one obtains:

$$E[y] \cong m(\underline{x}^0) + \frac{1}{2}\sum_{i=1}^{n}\sum_{j=1}^{n}cov[x_i, x_j]\left(\frac{\partial^2 y}{\partial x_i \partial x_j}\right)_{\underline{x}^0} \quad (7.4)$$

where *cov* is the covariance. Note that only if the model depends linearly on the inputs, the mean value of y can be calculated by evaluating the model in correspondence of the mean (nominal, best estimate) values of the input parameters; otherwise, one must account also for the covariance term (at least).

Typically, in practice the Taylor expansion is stopped at its first order, so that the expected value indeed becomes:

$$E[y] \cong y^0 = m(\underline{x}^0) \tag{7.5}$$

Under this assumption, the computation of the first order approximation of the output variance $Var[y]$ is straightforward. From (7.3) stopped at the linear term, one obtains:

$$Var[y] = E[(y - y^0)^2] \cong \sum_{i=1}^{n} Var[x_i] \left(\frac{\partial y}{\partial x_i}\right)_{\underline{x}^0}^2 + 2 \sum_{i=1}^{n} \sum_{j=i+1}^{n} \text{cov}[x_i, x_j] \left(\frac{\partial y}{\partial x_i}\right)_{\underline{x}^0} \left(\frac{\partial y}{\partial x_j}\right)_{\underline{x}^0} \tag{7.6}$$

If the input parameters are independent, the covariances in the second term are zero and a measure of the importance of the input parameter x_i with respect to the sensitivity of the output y to its variations can be defined as:

$$U_M(x_i) = Std[x_i] \left(\frac{\partial y}{\partial x_i}\right)_{\underline{x}^0} \tag{7.7}$$

where $Std[x_i]$ is the standard deviation of the uncertainty distribution of x_i. The method of moments is frequently used in practice for it is conceptually simple. Nevertheless, one must keep in mind that the expressions behind the calculation of the importance measure (7.7) are approximated, basically amounting to substituting the real model response function $m(\underline{x})$ with an hyper-plane tangent to it at the nominal values point, $\underline{x}^0$. Obviously, the (approximated) information deriving from this analysis is local and its validity is limited around $\underline{x}^0$. Finally, one should not underestimate the high computational costs of calculating the partial derivatives of complex models [3].

7.4 Discrete methods

7.4.1   Sensitivity on the nominal range

The effects on the model output y of variations in the values of the model inputs can be computed by varying each input parameter x_i from its minimum x_i^- to the maximum x_i^+ values, in the plausible variability range. One single parameter x_i at a time is varied, while the others $x_{j\neq i}$ are kept at their respective nominal values x_j^0 (Fig. 7.2). A measure of the importance of the input parameter x_i in influencing the model outut values, in the sense of sensitivity analysis, can then be defined as the difference of the output values obtained in correspondence of the two extrema x_i^- and x_i^+ of the input parameter [6]:

$$U_R(x_i) = m(x_i^+, x_{j\neq i}^0) - m(x_i^-, x_{j\neq i}^0) \tag{7.8}$$

Fig. 7.2 The method of sensitivity on the nominal range [6]

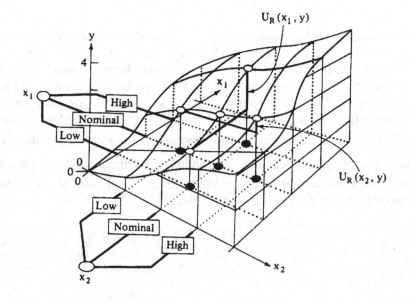

The index thereby defined is somewhat more than local since it evaluates the model in correspondence of the extreme values of each input parameter, but it is less than global since the parameters are considered one at a time. Nevertheless, it appears clear that the evaluation of the model only at the extreme values does not allow a detailed analysis of the trend of the function in the whole input parameter variability space.

To partially overcome these limitations, the method can be extended by evaluating the output (7.1) in correspondence of several values x_i within the parameter variability range, and not only at the extrema (parametric analysis), and by accounting for the joint effect of the simultaneous variation of more than one input parameter at the same time (Fig. 7.3).

Fig. 7.3 Joint parametric analysis for a model $y=m(x_1, x_2)$ [6]

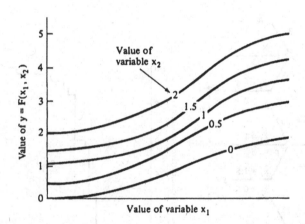

7.4.2 Event and probability tree

This method analyses systematically the effects of the combinations of different discrete values of the uncertain inputs. The variability ranges of the uncertain inputs are discretized in *levels* and the combinations of the levels of the various parameters are organized in a tree structure (Fig. 7.4) in which the nodes represent the uncertain parameters, while the branches starting from a node represent the possible values (levels) of the correspondent uncertain parameter. A sequence of connected branches generates an *event* whose output value y is obtained by evaluating the model with the parameters values of the branches on the event sequence. The uncertainty on the values of the parameters is quantitatively accounted for by assigning discrete probability values to the various branches of the tree. These probabilities are derived from the corresponding probability density functions (pdfs), conditioned on the values of the variables of the previous branches in the sequence. The probability $p(y)$ of the output value y associated to a given event sequence is computed by multiplying the probabilities of the single branches constituting the sequence. The set of couples $\{y,p(y)\}$ provides a discrete representation of the output distribution $f_Y(y)$ or the corresponding cumulative distribution function (cdf) $F_Y(y)$ (Fig. 7.4).

Fig. 7.4 Event probability tree [6]

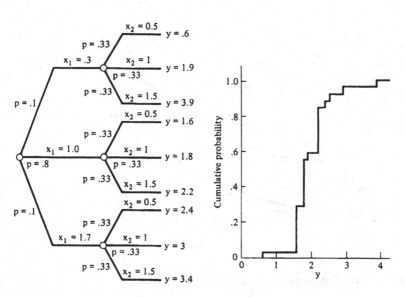

The main limitations of this approach consist in:

(1) the low number of levels which can be practically considered in the analysis, in order to avoid the combinatorial explosion of the event sequences to be evaluated

(2) the possible dependence of the results on the order in which the input variables are organized in the tree.

7.4.3 Discrete probability method

The discrete probability method allows limiting the combinatorial explosion of the event and probability tree [6,12,13]. Let us consider a simple model $y=m(x,z)$ which depends only on the two uncertain variables $x \sim f_X(x)$ and $z \sim f_Z(z)$. The pdf's $f_X(x)$ and $f_Z(z)$ can be discretized in n, s intervals of amplitude Δ_i and Δ_j, respectively, obtaining two sets of pairs $<x_i, p_i>$, $<z_j, q_j>$, $i=1,2,\ldots,n$, $j=1,2,\ldots,s$, where:

$$p_i = \int_{\Delta_i} f_X(x)dx \qquad x_i = \frac{1}{p_i}\int_{\Delta_i} x f_X dx \qquad i=1,2,\ldots,n$$

$$q_j = \int_{\Delta_j} f_Z(z)dz \qquad z_j = \frac{1}{q_j}\int_{\Delta_j} z f_Z dz \qquad j=1,2,\ldots,s$$

$$(7.9)$$

Notice that the discretization is performed in a way to preserve the total probability and the mean value at each interval.

Analogously to the event and probability tree method, the matrix $n\times s$ of the output values $y_{ij}=m(x_i,z_j)$ is built. Each output value is then assigned the probability $r_{ij}=p_i \cdot q_j$.

The discrete distribution of y is then condensed by defining a number of intervals $l=1,2,\ldots,t \ll ns$ and setting:

$$r_l = \sum_{y_{ij}\in l} r_{ij} \qquad y_l = \frac{1}{r_l}\sum_{y_{ij}\in l} r_{ij} y_{ij} \qquad l=1,2,\ldots,t \qquad (7.10)$$

The pairs $<y_l, r_l>$ define the discrete, "condensed" distribution of the uncertain variable y to be further manipulated depending on the purpose of the analysis. By doing so, the combinatorial explosion due to the effects of successive combinations of uncertain model parameters along the possible event sequences is contained.

Notice that after condensing the output pdf by (7.10), the possibly existing correlations among variables dependent on common parameters may be lost. Furthermore, the information on the contribution to the uncertainty by the single input parameters (i.e. the importance of the various uncertain parameters in a sensitivity analysis sense) is lost.

7.5 Monte Carlo method

A most comprehensive method for uncertainty and sensitivity analysis amounts to drawing a random sample of the uncertain input parameters values by Monte Carlo [3,6], computing the model output y for each value sampled and statistically manipulating the resulting sample of values. This way of proceeding obviously simplifies the problem formalization since it does not involve the systematic creation of a tree, which is instead implicitly generated through the random sampling, nor the discretization of the ranges of variability of the uncertain inputs.

The procedure consists in drawing from the assigned pdfs $f_{X_i}(x_i)$, $i=1,2,...,n$, a sequence of s realizations of each of the n input parameters:

$$\underline{x}_j = [x_{j1}, x_{j2},..., x_{jn}] \quad j=1,2,...,s \quad (7.11)$$

In correspondence of each of the s independently generated vectors $\underline{x}_j$, $j=1,2,...,s$, of n input values the model is evaluated, thus obtaining a sequence of output values:

$$y_j = m(x_{j1}, x_{j2},..., x_{jn}) \quad j=1,2,...,s \quad (7.12)$$

Such sequence constitutes an independent random sample of size s of the distribution of the (unknown) uncertainty of the output y. The sequences $\{x_j, y_j, j=1,2,...,s\}$ can be analyzed using classic statistical techniques for uncertainty and sensitivity analysis (e.g., linear regression techniques and variance decomposition techniques, see the following sub-chapters 7.6 and 7.7).

The Monte Carlo is a global method, covering the input parameter uncertainty space by one random sample. It is a simple method which provides a direct estimation of the output uncertainty distribution; it is a statistical approximated method, whose accuracy can be easily assessed (contrary to discrete methods) and improved by increasing the dimension s of the sample, but at the expense of increasing also the computational cost associated to the repeated model evaluations. Indeed, the global computational cost depends on the required degree of accuracy and for a given degree it increases linearly (and not exponentially) with the number of uncertain parameters n.

7.6 Linear regression method

Many of the commonly adopted methods for uncertainty and sensitivity analysis are based on linear regression techniques [4,11]. The importance of such methods derives from the possibility of efficiently obtaining estimates of the involved quantities, by introducing the hypothesis of model linearity. Such hypothesis allows achieving satisfactory estimates using limited size statistical samples. Nevertheless, the results of such an analysis may be misleading if the model structure is far from being linear.

Let us consider the approximated model:

$$y = m(\underline{x}) = y*(\underline{x}) + e = \underline{x}^T \underline{\beta} + e \qquad (7.13)$$

where $\underline{x}^T$ is the transposed vector of the variable $\underline{x}$, $\underline{\beta} = [\beta_1, \beta_2, \ldots, \beta_n]^T$ is the vector of the unknown coefficients of the model linear expansion and e is an error term which quantifies the deviation of the real model from its linear approximation; such deviation is usually treated as a random variable independent from the input parameters $\underline{x}$ and with zero average.

It is now possible to evaluate how the uncertainty propagates from the model inputs to its output by referring to the variance of the output distribution, which in the linear approximation can be expressed as follows:

$$Var[y] = Var[\underline{x}^T\underline{\beta}] + Var[e] \qquad (7.14)$$

If the input parameters are independent, one obtains

$$Var[y] = \underline{\beta}^T Var[\underline{x}]\,\underline{\beta} + Var[e] \qquad (7.15)$$

which can be expanded as

$$Var[y] = \sum_{i=1}^{n} \beta_i^2 Var[x_i] + Var[e] \qquad (7.16)$$

Assuming that the linear model (7.13) is a good approximation of the true model, the variance $Var[e]$ of the zero-average uncertain error, is expected to be small; then, the contribution of input parameter x_i to the output variance $Var[y]$ is identified in $\beta_i Var[x_i]$ and a reasonable representation of the importance of the input parameter x_i is the quantity:

$$R_i^2 = \frac{\beta_i^2 Var[x_i]}{Var[y]} \qquad i=1,2,\ldots,n \qquad (7.17)$$

In practice, the procedure of estimation of R_i^2, $i=1,2,\ldots,n$, proceeds as follows:

1. Sample a value ξ_1 from the distribution of x_1.
2. Sample all the remaining parameters x_i with $i \neq 1$, from the respective distributions $f_{x_i}(x_i)$, $i=1,2,\ldots,n$,. Including the value ξ_1, we obtain a realization of $\underline{x}$.
3. Insert $\underline{x}$ in the model $m(\underline{x})$ and compute y.
4. Repeat steps 2. and 3. r times (with the same value for ξ_1 obtained in step 1.). At the end of the k-th iteration, we obtain the value $y(1|k, \xi_1)$ and the vector of order n $\underline{x}(1|k, \xi_1)$, in which the first component is always equal to ξ_1, for any $k=1,2,\ldots,r$.
5. Compute the mean value

$$y(1|\xi_1) = \frac{1}{r}\sum_{k=1}^{r} y(1|k,\xi_1) \qquad (7.18)$$

6. Repeat steps 1. to 5. s times (with s different values for ξ_1). At the j-th iteration, we obtain the values $y(1|k, \xi_1^j)$ (end of step 4.) and $y(1|\xi_1^j)$, $j=1,2,\ldots,s$ (end of step 5.).
7. Compute the mean values

$$x(1) = \frac{1}{s}\sum_{j=1}^{s}\xi_1^j \qquad y(1) = \frac{1}{s}\sum_{j=1}^{s} y(1|\xi_1^j) \qquad (7.19)$$

8. Repeat steps 1. to 7. for the remaining parameters x_i, $i=2,3,\ldots,n$.
9. We have obtained the numerical values

$$y(i\,|\,k,\xi_i^j)$$
$$y(i\,|\,\xi_i^j) \qquad i=1,2,...,n; \quad k=1,2,...,r; \quad j=1,2,...,s \qquad (7.20)$$
$$x(i), y(i)$$

10. Based on these values, compute the statistical estimates $\overline{y}, \hat{V}[y], \hat{V}[x(i)], \hat{\beta}_i$ of $E[y], Var[y], Var[x_i]$ and β_i, respectively.

$$\overline{y} = \frac{1}{n}\sum_{i=1}^{n} y(i)$$

$$\hat{V}[y] = \frac{1}{nrs}\sum_{i=1}^{n}\sum_{k=1}^{r}\sum_{j=1}^{s}[y(i\,|\,k,\xi_i^j) - \overline{y}]^2$$

$$\hat{V}[x(i)] = \frac{1}{s}\sum_{j=1}^{s}[\xi_i^j - x(i)]^2 \qquad (7.21)$$

$$\hat{\beta}_i = \frac{\displaystyle\sum_{j=1}^{s}[y(i\,|\,k,\xi_i^j) - \overline{y}][\xi_i^j - x(i)]}{\displaystyle\sum_{j=1}^{s}[\xi_i^j - x(i)]^2}$$

from the least-square minimization of the error term in (7.13).

11. Finally, estimate the importance indexes

$$\hat{R}_i^2 = \frac{\hat{\beta}_i^2 \hat{V}[x(i)]}{\hat{V}[y]} \qquad i=1,2,...,n \qquad (7.22)$$

The two main limitations of the method are:

(1) the introduction of a bias in the estimation of both the mean and the variance of the output y
(2) the dependence on the hypothesis of model linearity.

The first limitation can be quantified and controlled, e.g. with the sample size s; the second does not allow the use of the method for non-linear models.

7.7 The variance decomposition method

For ease of explanation, let us consider a generic model whose output value y is dependent on only two input uncertain parameters (denoted by capital letters, for clarity of notation):

$$y = m(X_1, X_2) \tag{7.23}$$

No particular hypothesis is made on the model structure. The dependency of the output y on one variable, e.g. X_1, is approximated by the expected value of y computed with respect to the other input parameter X_2, conditioned by having set X_1 equal to a given numerical value x_1:

$$y*(x_1) = E_{X_2}(y \mid x_1) = \int y f(x_2 \mid x_1) dx_2 \tag{7.24}$$

Notice that by setting the variable X_1 equal to the numerical value x_1, y depends only on the variable X_2.

Like for the linear regression method, the uncertainty propagation from input to output is evaluated in terms of the variance of the output distribution $Var[y]$, which can be decomposed in the following way [1,4,5,14] (for further details, see the Appendix at the end of the chapter):

$$Var[y] = Var_{X_1}[E_{X_2}(y \mid x_1)] + E_{X1}[Var_{X_2}(y \mid x_1)] \tag{7.25}$$

The importance associated to parameter X_1 is related to its average contribution to the variance of y, i.e. $Var_{X_1}\left[E_{X_2}(y \mid x_1)\right]$. The second term of the decomposition in (7.25) represents instead the variability of y which is not dependent on X_1 but on the other variable X_2: such variability must be small if X_1 is "important". Therefore, it seems reasonable to retain as importance index for the variable X_1 the quantity:

$$\eta_1^2 = \frac{Var_{X_1}[E_{X_2}(y \mid x_1)]}{Var[y]} \tag{7.26}$$

The procedure for estimating η_1^2 is as follows [14]:

1. Sample s random values of x_1, $\{x_1^1, x_1^2, ..., x_1^s\}$.

2. For each x_1^j, sample r values x_2^k, $k = 1, 2, ..., r$ from the conditioned distribution $f_{X_2|X_1}(x_2 | x_1^j)$, $j = 1, 2, ..., s$.

3. Evaluate the r output values $y^{jk} = m(x_1^j, x_2^k)$, $j = 1, 2, ..., s$, $r = 1, 2, ..., k$. Such values constitute the elements of the output matrix of order (s, r).

4. For each row $j = 1, 2, ..., s$ of the matrix, compute

$$\hat{y}*(x_1^j) = \frac{1}{r}\sum_{k=1}^{r} y^{jk} \cong E_{X_2}[y | x_1^j] \qquad j = 1, 2, ..., s \qquad (7.27)$$

5. Compute the estimate $\overline{y}$ of the expected value of $E[Y]$

$$\overline{y} = \frac{1}{s}\sum_{j=1}^{m} \hat{y}*(x_1^j) \cong E[y] \qquad (7.28)$$

Compute the estimates $\hat{V}_{X_1}[E_{X_2}(y | x_1)]$ and $\hat{V}[y]$ of the variances $Var_{X_1}[E_{X_2}(y | x_1)]$ and $Var[y]$, respectively

$$\hat{V}_{X_1}[E_{X_2}(y | x_1)] = \frac{1}{s-1}\sum_{j=1}^{s}[\hat{y}*(x_1^j) - \overline{y}]^2$$

$$\hat{V}[y] = \frac{1}{sr-1}\sum_{j=1}^{s}\sum_{k=1}^{r}(y^{jk} - \overline{y})^2 \qquad (7.29)$$

6. Evaluate the importance indexes

$$\hat{\eta}_1^2 = \frac{\hat{V}_{X_1}[E_{X_2}(y | x_1)]}{\hat{V}[y]} \qquad (7.30)$$

Theoretically, this global method has the advantage of avoiding the introduction of hypotheses which would constrain and limit the model structure. Nevertheless, it entails a higher computational cost (e.g., more than that required by linear regression analysis), which obviously depends on the model complexity.

Furthermore, notice that the method of linear regression previously illustrated in sub-chapter 7.6 represents a particular case of the more general variance decomposition method under an opportune linearity hypothesis on the approximating model.

It is also possible to verify that $R^2 \leq \eta^2$, which means that a linear regression analysis performed on models characterized by a strong non-linearity might fail to highlight inputs whose contribution to the output uncertainty is instead important.

Example 7.1 [15]

The example considered concerns the variance decomposition-based sensitivity analysis of the reliability model of a system in which the values of the components failure rates are uniformly uncertain within given ranges of variability extending one order of magnitude from given base case values.

The model of the system is realistically complicated by accounting for aging, maintenance and stand-by logics; hence, its analytical evaluation is impractical and resorting to Monte Carlo simulation (chapter 2) is required. However, this is excessively burdensome in terms of computation time, since the variance decomposition sensitivity analysis method requires the repetition of a large number of system evaluations, each one to be performed by Monte Carlo. This problem is here circumvented by substituting the Monte Carlo simulation model with a fast, approximated model based on a neural network [16], which is appropriately trained on the results of a Monte Carlo evaluation of the system reliability model to quickly provide, with reasonable approximation, the values of the quantities of interest for the sensitivity analysis.

The type of neural network here employed is the classical multi-layered, feed-forward one trained by the error back-propagation method [16]. The networks used have been generated with a user-friendly software NEST (NEural Simulation Tool) developed at the Department of Energy of the Polytechnic of Milan (http://lasar.cesnef.polimi.it).

The training patterns for the neural network are constituted by the values of the uncertain system components failure rates, sampled within the respective ranges of variability, and by the system unreliability values evaluated by Monte Carlo simulation in correspondence of the sampled failure rates values. The Monte Carlo evaluation of the system unreliability for given values of component failure rates has been performed by means of a user-friendly Monte Carlo simulation code, MARA (Monte Carlo Availability Reliability Analysis), also developed at the same Department of the Polytechnic of Milan (http://lasar.cesnef.polimi.it). Several procedures have been implemented in the code which allow the definition of different logics of components' operation e.g. stand-by and load-sharing, the consideration of aging and maintenance, and other realistic aspects of the system behaviour. For the case study of interest, these are described in details in [15].

The system considered is constituted by three macro-components in series (Fig. 7.5). Each macro-component is composed of a redundancy configuration (called "block") of components with constant base values of failure and repair rates (Table 7.1).

Fig. 7.5 System layout

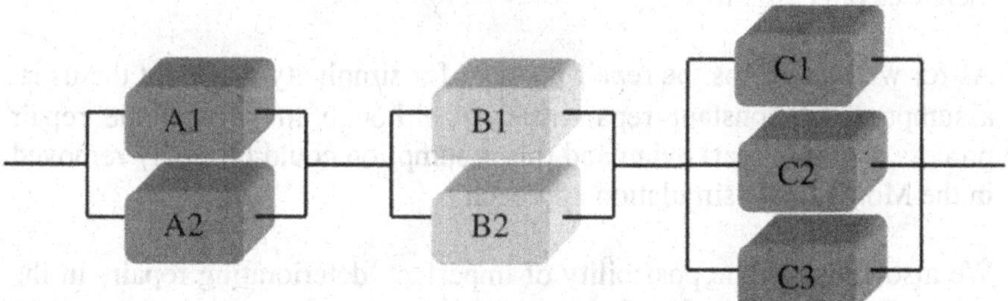

Table 7.1 Components' failure and repair rates

Component i	Failure rate λ_i [y^{-1}]	Repair rate μ_i [y^{-1}]
A1	$6.0 \cdot 10^{-3}$	$1.7 \cdot 10^{-1}$
A2	$2.6 \cdot 10^{-3}$	$1.0 \cdot 10^{-1}$
B1	$5.3 \cdot 10^{-3}$	$3.0 \cdot 10^{-1}$
B2	$3.6 \cdot 10^{-3}$	$1.0 \cdot 10^{-1}$
C1	$8.1 \cdot 10^{-3}$	$5.0 \cdot 10^{-1}$
C2	$5.3 \cdot 10^{-3}$	$3.0 \cdot 10^{-1}$
C3	$7.0 \cdot 10^{-3}$	$5.0 \cdot 10^{-1}$

Components are assumed to age in time according to a linear model for the failure rate $\lambda(t)$ [17],

$$\lambda(t) = \lambda_0 + a \cdot t \qquad (7.31)$$

where λ_0 is the constant base value and a is the aging rate.

The effects of aging are mitigated through preventive maintenance actions, performed with period τ, which rejuvenate the component (see below). The period τ is chosen sufficiently small that the failure rate $\lambda(t)$ increases only slightly.

As for what concerns the repair process, for simplicity we adopt the usual assumption of constant repair rate, μ, although in general the repair process is all but markovian and this assumption could be easily removed in the Monte Carlo simulation approach.

We also consider the possibility of imperfect, deteriorating repairs in the sense that as a result of a repair action the component does not necessarily return to an "as good as new" condition but may come out more fragile and prone to future failures. To account for imperfect, deteriorating repairs, we adopt a modified Brown-Proschan model of stochastic repairs which postulates that a system is repaired to an "as good as before" condition (*minimal repair*) only with a certain probability p and is, otherwise, returned in a "deteriorated" condition (*deteriorating repair*) [18-19]. Thus, these two conditions obey a Bernoulli distribution. In case of deteriorating repair, the component emerges with a failure rate increased and a repair rate reduced by given percentages, π_λ and π_μ respectively, with respect to the values before the failure. Hence the deterioration has a combined effect of increasing the failure rate of the component and of somewhat 'complicating' its repair which thus requires, on average, more time, as modeled by the reduced repair rate. The analyst-defined parameters π_λ and π_μ specify the amount of deterioration induced by the failure-repair process.

The maintenance model here proposed is a modification of a previously adopted model [20] and it is based on the following assumptions: i) the maintenance of a component is performed with variable period τ and the periodicity varies with the component's deterioration due to imperfect repair (Fig. 7.6); ii) the period τ between successive maintenances is made up by the time intervals during which the component is operating in its active state. In other words, when the system is inactive, e.g. in standby mode, the elapsed time does not contribute to τ; iii) maintenance actions are such to restore the conditions existing at the beginning of the previous maintenance period; iv) maintenance actions are instantaneous and a sufficient number of maintenance teams is available at all times for all kinds of maintenance required.

In realistic situations, the maintenance activities become more and more frequent as the component ages and its reliability characteristics deteriorate. In our model, deterioration of a component occurs due to the imperfect repairs, in accordance to the Brown-Proschan model previously introduced, and we allow for an adaptive schedule of maintenance intervention according to which the ratio between the maintenance period τ and the mean time $1/\lambda$ between successive failures is kept constant. Thus, after a minimal repair $\lambda_{new} = \lambda_{old}$ and, thus, $\tau_{new} = \tau_{old}$; on the contrary, after a deteriorating repair, $\lambda_{new} = (1+\pi_\lambda)\lambda_{old}$ and, then, $\tau_{new} = \tau_{old}$ $\lambda_{old} / \lambda_{new} = \tau_{old} / (1+\pi_\lambda)$. Fig. 7.7 shows an example with $\pi_\lambda=1$.

Finally, it seems reasonable to assume that for each component the period between maintenances is a fraction $1/\alpha_\tau$ of its MTTF, i.e.:

$$\tau = \frac{1}{\alpha_\tau}\bar{t_f} = \frac{1}{\alpha_\tau}\frac{1}{\lambda^*(\tau)} = \frac{1}{\alpha_\tau}\ \frac{1}{\lambda_0 + a\dfrac{\tau}{2}} \tag{7.32}$$

We then obtain the explicit expression for the maintenance period:

$$\tau = \frac{\lambda_0}{a}\left[\sqrt{1 + \frac{1}{\alpha_\tau \frac{\lambda_0^2}{2a}}} - 1\right] \tag{7.33}$$

By so doing, the maintenance period τ of each component is linked to the values characterizing its linear degradation and failure behaviors.

Fig. 7.6 Linearly aging failure rate and mitigating effect of maintenance

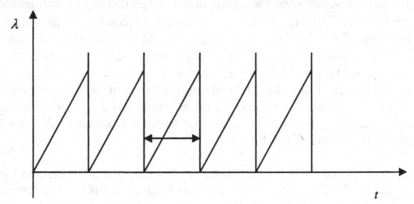

Fig. 7.7 Adaptive maintenance period for a linearly aging component. After component failure and repair (lasting a time T_{rep}) the component ages according to the Brown-Proschan model and the maintenance period is shortened from τ_{old} to τ_{new}

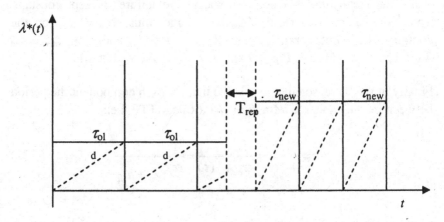

Finally, in the Monte Carlo simulation model, the active-to-standby and standby-to-active transitions and the load-sharing dependences are dealt with by establishing a proper correlation between the system states before and after the transition, following the transition rate multiplicative correlation model proposed in [21]. This model allows one to modify, upon occurrence of a system state change, the dependent transition rates by multiplying the original values times a pre-defined multiplication factor which depends on the current and arrival system configurations.

With regards to the system of Fig. 7.5, the following specific assumptions are considered:

- components B2, C2, C3 are in cold stand-by;
- imperfect repairs are possible according to the previously introduced Brown-Proschan model with a probability $p = 0.1$. Such imperfect repairs have a deteriorating effect in that they determine an increase by a factor of $1+\pi$ of the failure rate of the repaired component and a corresponding reduction of the repair rate of the same factor, where $\pi = 0.1$.
- the components are subject to maintenance at variable periods τ as determined from the model previously illustrated.

The system model is complicated enough so as to entail a Monte Carlo simulation evaluation. Embedding such simulation procedure within a variance decomposition scheme of sensitivity analysis can lead to impractically large computing times. On the contrary, using the trained neural network for the model evaluations there is a ratio of 1/50,000 in terms of computation speed with respect to the Monte Carlo simulations.

The neural network considered is made up of 7 input nodes for the 7 components' failure rates and 1 output node for the system unreliability at mission time. The training set is composed of 500 different input/output patterns generated by the MARA code, with failure rates values uniformly sampled by crude Monte Carlo within the ranges of Table 7.2. Note that the components A_1, A_2 and B_2 of blocks A and B have been purposely chosen to have nominally the same failure rates so as to allow us to highlight the effects of different logic configurations.

The neural network training has been performed with the usual batch back-propagation algorithm [16] with the following parameters:

- learning rate, momentum: 0.5 (optimized), 0.6 (optimized)
- number of hidden nodes, n_h : 7 (optimized)
- patterns per batch, n_p : 20
- batches, n_b : 100,000
- repetitions   : 10

The strategy employed to find the optimal values of learning rate, momentum and number of hidden nodes has consisted simply in an automatic trial and error search over a pre-defined cubic mesh of those parameters. The optimal training strategy is that which gives the minimum cumulative quadratic error between the true and network-estimated values of unreliability on all the patterns of the training set (the interested reader may consult [15] for further details on the network training procedure).

Table 7.2 Range of variability of failure and repair rates, obtained with a multiplication factor of 1/3.16 and 3.16

Component i	Nominal State	Failure rate λ_i [y^{-1}]			
		Nominal value	Lower limit	Upper limit	Mean value
A1	Working	$6.00 \cdot 10^{-3}$	$1.90 \cdot 10^{-3}$	$1.90 \cdot 10^{-2}$	$1.04 \cdot 10^{-2}$
A2	Working	$6.00 \cdot 10^{-3}$	$1.90 \cdot 10^{-3}$	$1.90 \cdot 10^{-2}$	$1.04 \cdot 10^{-2}$
B1	Working	$6.00 \cdot 10^{-3}$	$1.90 \cdot 10^{-3}$	$1.90 \cdot 10^{-2}$	$1.04 \cdot 10^{-2}$
B2	Stand-by	$1.50 \cdot 10^{-2}$	$4.75 \cdot 10^{-3}$	$4.74 \cdot 10^{-2}$	$2.61 \cdot 10^{-2}$
C1	Working	$8.10 \cdot 10^{-3}$	$2.56 \cdot 10^{-3}$	$2.56 \cdot 10^{-2}$	$1.41 \cdot 10^{-2}$
C2	Stand-by	$2.50 \cdot 10^{-2}$	$7.91 \cdot 10^{-3}$	$7.90 \cdot 10^{-2}$	$4.35 \cdot 10^{-2}$
C3	Stand-by	$5.00 \cdot 10^{-2}$	$1.58 \cdot 10^{-2}$	$1.58 \cdot 10^{-1}$	$8.69 \cdot 10^{-2}$

Fig. **7.8** depicts the network architecture with the 7 input nodes receiving the 7 component failure rates values, the 7 hidden nodes performing intermediate processing and the single output node providing an estimate of the system unreliability at mission time.

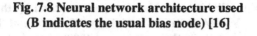

**Fig. 7.8 Neural network architecture used
(B indicates the usual bias node) [16]**

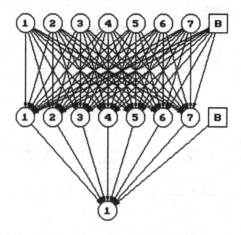

The results of the optimized training phase for the neural network of Fig. 7.8, trained to estimate the system unreliability at mission time, are presented in Fig. 7.9. In particular, Fig. 7.9 shows the comparison of the estimates of the system unreliability at mission time obtained by the MARA code with 10^6 trials (ordinates) with those obtained by the neural network (abscissas) for different values of the 7 input failure rates of the components sampled within their respective ranges. The top picture regards the estimates on the 500 examples of the training set whereas the bottom one regards the estimates on other 500 examples uniformly sampled by crude Monte Carlo within the ranges of Table 7.2 and never processed by the neural network during the training. The approximation errors made by the neural network are certainly satisfactory, considering the fact that the space spanned by the 7 input parameters is rather large.

Fig. 7.9 System unreliability at mission time computed by the Monte Carlo simulation code MARA with 10^6 trials (ordinates) and by the neural network (abscissas): top, in the training set (500 examples); bottom, in the test set (500 examples)

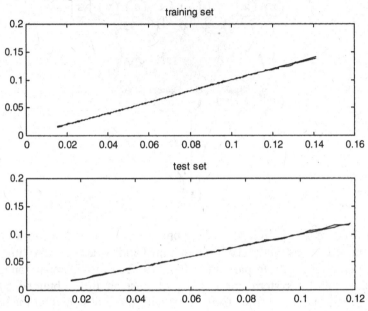

With the neural network trained over the ranges of Table 7.2, it is possible to perform a variance decomposition sensitivity analysis of the system model in various situations characterized by different nominal values of the input parameters, so as to put in evidence, for example, how a bad component affects the system behavior or which are the effects of a high grade of redundancy on the uncertainty of the top event.

In the following are reported the results of the calculation of the index of importance η_X^2 (computed according to eq. 7.26) for the failure rates of the 7 components taken one at a time (i.e., the vector X contains, in turn, the failure rate of each component, one at a time) and for the 3 blocks (A, B, C) that compose the system (i.e., the vector X contains, in turn, the failure rates of all the components in each block).

In Fig. 7.10 we report the indexes of importance of the individual failure rates with respect to system unreliability. We can observe that although component B1 has the same nominal failure rate value as A1 and A2, the presence of the companion component B2 in stand-by in block B reflects on the value of the index of importance. Indeed, as expected, when a component is supported by another in stand-by its uncertainty strains less on the uncertainty of the system unreliability at mission time than it would if it were supported by an identical parallel component in working state.

The effect of the logic configuration also shows in the low indexes values of the failure rates pertaining to the components of block C, where there is a higher grade of redundancy and where there are two components in cold stand by.

Fig. 7.10 Index of importance of the failure rate of a single component with respect to the system unreliability at mission time

Index of importance of failure rate λ on system unreliability at mission time	
Component	η^2
A1	0.257
A2	0.258
B1	0.208
B2	0.202
C1	$2.35 \cdot 10^{-2}$
C2	$1.92 \cdot 10^{-2}$
C3	$1.62 \cdot 10^{-2}$

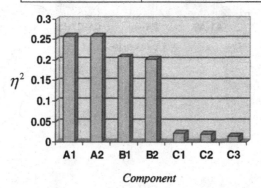

Component

Considering the blocks A, B, C (Fig. 7.11) their indexes of importance reflect the logic of the plant configuration so that uncertainties in the failure rates values pertaining to the least reliable block A have a greater impact on the uncertainty in the system unreliability than those pertaining to the more reliable blocks B (identical to A except that B_2 is in cold stand-by and not in a critical parallel) and C (with two cold stand-by components).

Fig. 7.11 Index of importance of the parameters of a block of components with respect to the system unreliability at the mission time

Index of importance of each block on system unreliability at mission time	
Block	η^2
A1 –A2	0.610
B1 – B2	0.356
C1 –.C2 – C3	$4.862 \cdot 10^{-2}$

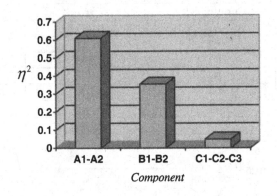

7.8 Sobol indexes and Fourier Amplitude Sensitivity Test

Concluding the review of the methods used to assess the output uncertainty and sensitivity resulting from uncertain inputs, let us briefly present the underlying principles of two more sophisticated techniques recently developed by the scientific community: *Sobol indexes* [3,7,10] and the *Fourier Amplitude Sensitivity Test (FAST)* [9,22].

Both methods allow for a global uncertainty analysis and are based on a decomposition of the variance in the contributions due to the single uncertain parameters and their combinations:

$$Var[y] = \sum_{i=1}^{n} D_i + \sum_{i=1}^{n} \sum_{\substack{j=1 \\ j \neq i}}^{n} D_{i_j} + ... \tag{7.1}$$

where, the generic term $D_{i_1,i_2,...,i_s}$ is the average contribution to the variance of y due to all the variables of the subset $\{x_{i_1}, x_{i_2}, ..., x_{i_s}\}$ considered varying at the same time. The importance measure of the subset of input parameters $\{x_{i_1}, x_{i_2}, ..., x_{i_s}\}$ is then defined as:

$$S_{i_1,i_2,...,i_s} = \frac{D_{i_1,i_2,...,i_s}}{D} \tag{7.2}$$

Notice that the first order indexes S_{i_j}, which measure the importance of the single input parameters, coincide with the importance indexes of the variance decomposition method illustrated in the previous sub-chapter 7.7.

The Sobol and FAST methods allow defining for each parameter an overall index of importance as the sum of the indexes (7.35) in which such parameter appears, by itself or in combination with others. For example, let us consider a three-parameter model $y=m(X_1,X_2,X_3)$, whose output variance can be decomposed in the following terms:

$$Var[y] = D_1 + D_2 + D_3 + D_{12} + D_{13} + D_{23} + D_{123} \qquad (7.3)$$

The overall importance index for X_1 is then

$$TS_1 = S_1 + S_{12} + S_{13} + S_{123} \qquad (7.4)$$

In Sobol's method, each partial index (7.35) is computed by evaluating a multi-dimensional integral using the Monte Carlo method. The advantage is that the overall effect due to one parameter, i.e. the overall importance index TS of (7.37), can be obtained with one single Monte Carlo integration.

The FAST technique amounts to introducing a search curve to "fill" appropriately the input parameters variability space. The integration for computing the various variance contributions is performed along this line and not within the multi-dimensional space. The model is then expanded in Fourier series to obtain the various contributions of the uncertain input parameters to the output variance.

The main practical limitations of these innovative methodologies are due to:

(1) a significant mathematical complexity
(2) the difficulty in finding appropriate search curves along which to perform the integration (FAST)
(3) the need of executing many model calculations to obtain the estimation of the parameters contributions to the variance.

7.9 Model structure uncertainty

As anticipated in the Introduction 7.1, a model provides a representation of a real system which is dependent on a certain number of hypotheses regarding its structure and parameters. The uncertainties in the model structure derive from the presence of different sets of hypotheses, each one consistent with the available data concerning the system under analysis, and from possible simplifications introduced both during the model conceptualization and its codification in a mathematical format; finally, further uncertainties may arise from discretizations and simplifications introduced during the coding of the mathematical model or from programming errors [23-29]. Representing such uncertainties is an extremely difficult problem and, nowadays, a satisfactory and generally accepted methodology has yet not been found.

The two most common approaches generally followed in practice to analyse model uncertainties are the *alternative models* [30-36] and the *adjustment factor* approaches [30-39], also called the model set and prediction expansion approaches, respectively [40].

7.9.1   The alternative models approach

Let us suppose that there exists a set of M plausible models m_i, $i=1,2,...,M$, each characterized by a set of uncertain parameters $\underline{a}_i$ with distribution $\pi(\underline{a}_i | m_i)$. Given certain values of the independent variables $\underline{x}$ and parameters $\underline{a}_i$, each model will return a different result, $y_i = m_i(\underline{x}, \underline{a}_i)$. The model uncertainty is quantified in terms of a discrete probability distribution $p(m_i)$, $i=1,2,...,M$, which somewhat reflects the subjective confidence in the validity of the model m_i for describing the phenomenon under analysis. For estimating the output y, one can resort to the total probability theorem, viz.:

$$y = \sum_{i=1}^{M} p(m_i) \int m_i(x_i, a_i) \pi_i(a_i | m_i) da_i \qquad (7.5)$$

Such method is undoubtedly intuitive and consistent with the Bayesian approach of probabilistic analysis. Nevertheless, difficulties arise in defining the meaning of the quantity $p(m_i)$, since it is not clear the significance of assigning to a model the probability of being the true one or of ensuring the "best representation", etc. Furthermore, the approach is based on the hypotheses of exclusiveness and exhaustivity of the models which are both hard to verify in practice. In fact, concerning the model exclusiveness, often the plausible models considered are based mostly on the same hypotheses and differ only for few hypotheses related to the treatment of specific phenomena; concerning the model exhaustivity, it seems difficult to ensure that all the possible models for a phenomenon have been considered, including the true one which must be within the exhaustive set: in fact, in the ideal situation of complete knowledge of the system under study, the Bayesian updating should lead the probability distribution $p(m_i)$ to become fully concentrated (with probability equal to one) on the true model.

Despite these difficulties, this approach has been applied in the epidemiologic [26] and hydro-geologic fields [24,25,27].

7.9.2 Adjustment factor approach

According to this approach, one single model $m*$ is considered, typically the most plausible, and an adjustment factor is directly applied to the model output to account for the associated uncertainties. More specifically, such perturbation is a sort of error term with which one accounts for the structural uncertainties in the model due to the incomplete knowledge of the phenomena. In practice, one adopts additive or multiplicative adjustment factors, viz.:

$$y = m*(\underline{x},\underline{a}*) + D_a* \qquad y = m*(\underline{x},\underline{a}*) \cdot D_m* \qquad (7.6)$$

The adjustments D_a* and D_m* are uncertain and described in terms of appropriate probability distributions.

The approach is obviously empirical and encounters many difficulties in determining the probability distribution representing the uncertainty in the adjustment of the reference model output. Despite these difficulties, the approach has been applied in seismic [29] and fire risk models [30].

Example 7.2 [35]

The example concerns the evaluation of alternative conceptual models for groundwater flow and solute transport at a high level waste repository site in unsaturated, fractured tuff formations. In particular, the six conceptual models presented in [24] are examined with the objective of quantitatively assessing the uncertainty associated to their predictions, within the frameworks of model uncertainty analysis presented in sub-chapters 7.9.1 and 7.9.2. A brief description of the six models is first provided. For further details on the specific models, the reader should consult [25].

The six models considered are based on a set of common assumptions and simplifications but differ by some fundamental hypotheses on the flow and transport mechanisms in the groundwater system. In particular, four different models for the groundwater flow have been considered, and when these are combined with the assumptions for the solute transport, they give rise overall to the six different models. All of the models are based on the same system geometry. A spatially uniformly distributed flux at the repository horizon equal to 0.1 mm/yr is assumed as boundary condition for all models. Steady-state groundwater flow through a one-dimensional system, occurring from the base of the repository to the water table, was also assumed. Fifteen hydrogeologic layers in the vertical direction were used to describe the hydraulic and geologic properties in five of the six models. The sixth model was based on an equivalent four-layer stratigraphy based on average properties. Only four radionuclide species, Tc-99, I-129, Cs-135 and Np-237, are assumed to be released from the repository and transported as solute. The transport model is based on the assumption of a single, dominant, non-branching transport path. Radionuclide retardation factors were based on the distribution coefficient K_d for each radionuclide. No other chemical reactions were included. Gaseous phase transport was not included either. Table 7.3 reports the principal characteristics which distinguish the models.

Table 7.3 Characteristics of the six alternative models [25]

Model	Distinguishing Assumptions
1	Single-continuum; groundwater flow and solute transport occur only in the porous matrix
2	Single-continuum; groundwater flow and solute transport occur only in connected fractures; no retardation effects in the fracture continuum
3	Single-continuum; groundwater flow as in 2; solute transport through the fractures and also through the immobile matrix water by diffusion; retardation effects only in the matrix by sorption
4	Dual-continuum; simultaneous matrix and fracture flow; solute transport occurs in the medium with highest pore velocity; retardation effects only in the matrix by sorption
5	Dual-continuum; groundwater flow as in 4; transport described using an equivalent porous medium approximation; retardation effects only in the matrix by sorption
6	Dual-continuum; groundwater flow and solute transport as in 4; only four equivalent hydrogeologic layers

Table 7.4 reports the uncertainty stemming from the six groundwater flow and contaminant transport models, in terms of the cumulative release to water table in curies, for a given set of values of the parameters. The cumulative release to the water table for the all-fracture model (model 2) is, for all practical purposes, equal to the cumulative release from the source, and this is true for all radionuclides. This is because of the relatively short travel time for this case, which results from the relatively high fracture velocities and lack of radionuclide retardation in the fracture continuum. Also, because of the very short travel time, a negligible amount of the inventory is lost to radioactive decay. Adding any matrix transport and/or matrix retardation significantly decreases the cumulative release. In fact, for Cs-135 and Np-237, predictions of any of the models that contain matrix transport result in zero release to the water table, primarily because of the high matrix retardation factors for these two isotopes.

Table 7.4 Cumulative release to water table in curies [25]

MODEL	Tc-99	I-129	Cs-135	Np-237
1	197	7.15	0.0	0.0
2	$6.26 \cdot 10^5$	1510	$1.68 \cdot 10^5$	0.043
3	$5.8 \cdot 10^{-4}$	$8.5 \cdot 10^{-7}$	$4.5 \cdot 10^{-12}$	$4.3 \cdot 10^{-15}$
4	$5.29 \cdot 10^{-4}$	515	0.0	0.0
5	$5.52 \cdot 10^{-4}$	416	0.0	0.0
6	$3.41 \cdot 10^{-4}$	535	0.0	0.0

Let us begin the analysis by first looking at the models of groundwater flow available. A close examination of the models from this point of view shows that there are actually only four different structures, as given in Table 7.5 where the superscript f stands for *flow*.

Table 7.5 Groundwater flow models

Flow Structure	Model (Table 7.3)	Characteristics
S_1^f	1	Groundwater flow in porous matrix only
S_2^f	2, 3	Groundwater flow in fractures only
S_3^f	4, 5	Groundwater flow in matrix and fracture simultaneously
S_4^f	6	Groundwater flow in matrix and fracture simultaneously, but four stratigraphic layers only

Fig. 7.12 reports the deterministic predictions of the head distribution in the system provided by the four models for a given set of parameter values. In the analysis that follows we ignore the effects due to uncertainty in the parameters and we focus on the uncertainty due to the model structure.

**Fig. 7.12 Uncertainty in the head distribution due to the four different
models of groundwater flow**

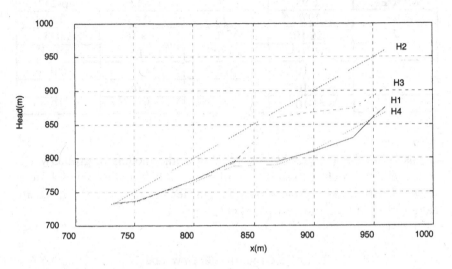

Following the alternate-hypotheses approach introduced in sub-chapter
7.9.1, suppose that an expert considers S_3^f as the most plausible model S^*
and assigns a value of 0.6 to its probability, meaning that given the
available information, he or she feels 60% confident that this model
would provide an appropriate description of the groundwater flow
phenomenon for the objective of the analysis. The other structures S_1^f,
S_2^f, S_4^f are seen to differ from S_3^f by only one assumption on the
fundamental mechanism of flow. On the basis of the available evidence,
the expert assigns values to the probabilities of the alternate hypotheses.
An example is reported in Table 7.6 where for example by assigning a
value of 0.01 to the probability of model 1 the expert expresses his or her
low confidence regarding the hypothesis of an all-matrix flow. Also, for
the sake of this example, we assume that the four models form an
exhaustive set, in the sense that they are expected to provide a
satisfactory representation of the retrospective truth and, therefore,
enough hedging against uncertainty.

Table 7.6 Assessment of model uncertainty for the groundwater flow models of Table 7.5: $S^* = S_3^f$

S_i^f	$p(S_i^f)$
S_1^f	0.01
S_2^f	0.04
S_4^f	0.35

Fig. 7.13 shows a comparison between the prediction provided by the best model S_3^f and the Bayesian estimator obtained by introducing the probability values of Table 7.6 into eq. (7.38), which provides a more appropriate representation accounting for the uncertainty in the model structure.

Fig. 7.13 Comparison of the head distributions given by the best model $S^* = S_3^f$ (solid) and by the Bayesian estimator of (7.38) with the probabilities of Table 7.6 (dashed)

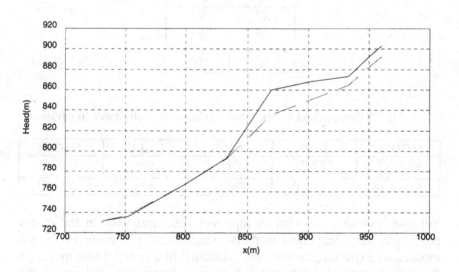

To complete the example, we now consider all six models for groundwater flow and contaminant transport, which we denote as S_i^{ft}, i=1, 2, ... ,6, in the order given in Table 7.3, and assume that the expert chooses $S_4^{ft} = S^*$ with $p(S^*)=0.45$. Model structures S_1^{ft}, S_2^{ft}, S_3^{ft} differ by two hypotheses while S_5^{ft} and S_6^{ft} only by one. The corresponding probabilities are given in Table 7.7. The application of the probabilities in Table 7.7 as weights in the Bayesian estimator of (7.38) leads to a measure of cumulative release as indicated in Table 7.8 where it is compared to the values obtained by $S^* = S_4^{ft}$. Again, the estimates obtained by means of (7.38) give due count to the uncertainty inherent in the model structure. Notice how the estimates are quite close in all cases, as the mean is driven by models S_3^{ft} and S_4^{ft} which happen to have similar probabilities and predictions.

Table 7.7 Assessment of model uncertainty for the groundwater flow and contaminant transport models of Table 7.3: $S^* = S_4^{ft}$

S_i^{ft}	$p\left(S_4^{ft}\right)$
S_1^{ft}	0.01
S_2^{ft}	0.04
S_3^{ft}	0.1
S_5^{ft}	0.3
S_6^{ft}	0.1

Table 7.8 Comparison of cumulative releases to water table in curies

MODEL	Tc-99	I-129	Cs-135	Np-237
$S^* = S_4^{ft}$	$5.29 \cdot 10^4$	515	0.0	0.0
Bayesian	$6.88 \cdot 10^4$	470.52	672	$1.72 \cdot 10^{-3}$

Suppose now that we wish to represent the uncertainty in the output predictions regarding the cumulative release of I-129 given by the six models, by expanding directly the prediction of a selected best model S^* through a properly assessed adjustment factor D^*, following the framework of model uncertainty analysis presented in sub-chapter 7.9.2. To be consistent with the earlier analysis we assume that the expert chooses $S^* = S_4^{ft}$. Indeed, one could very well think of a two-step process which combines the two approaches proposed in sub-chapters 7.9.1 and 7.9.2: first, the expert considers the set of alternate models and selects the best model S^*; then, he or she proceeds to modify the predictions of S^* so as to account for its inherent uncertainty as indicated by the available information, including that provided by the identified alternate models. This process has the potential of leading to an efficient and explicit method for evaluating and propagating model uncertainty.

In our example, we assume that the expert uses the information given by the alternate models to build a distribution for the modification factor D^*, which then represents model-to-model uncertainty. In most practical cases, other information combined with expert judgments will also be used to arrive at the final distribution for D^*.

Notwithstanding the paucity of the data, for the purpose of this example, the residuals $(y_i - y^*)$ of the five alternate models can be equally weighed to produce a mean value $\overline{D}^* = -21.37$ and a standard deviation $\sigma_{E^*} = 616.66$, which indeed reflects the large uncertainty present in the predictions given by the models. The solid line in Fig. 7.14 represents the distribution of D^*, under a normality assumption. Taking into account the expert's judgments on the credibility of the various alternatives, expressed in terms of probabilities $p(S_i)$, we can modify the distribution just obtained by using the probabilities of Table 7.7, to estimate a Bayesian predictive distribution according to (7.38) (dotted line in Fig. 7.14). In this case, this leads to a slight shift of the curve towards more negative values and a reduction in the uncertainty, the new mean and standard deviation values being -44.48 and 262.32, respectively.

The results of the above analyses can then be presented to the expert for discussion and refinement. The expert is then left free to reconsider the previously assessed values and adjust them, if that is the case, on the basis of the analysis of the available evidence and of other considerations reflecting his/her personal expertise. The process continues in an iterative manner, with the goal of producing a final distribution which properly reflects the expert's beliefs.

Fig. 7.14 Comparison of the distributions of *D** obtained by simple(solid) and weighed averaging (dashed)

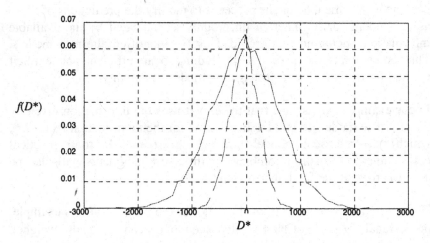

7.10 References

[1]   McKay, M. D., *Evaluating Prediction Uncertainty*, NUREG/CR-6311, LA-12915-MS, Los Alamos National Laboratory, U.S. Nuclear Regulatory Commission, 1995, p.5.

[2]   Saltelli, A., Ratto, M., Andres, T., Campolongo, F., *Global Sensitivity Analysis: The Primer Global Sensitivity Analysis: The Primer*, Wiley, Hardcover, 2008.

[3]   Helton, J. C., *Uncertainty and Sensitivity analysis Techniques for Use in Performance Assessment for Radioactive Waste Disposal*, Reliability Engineering and System Safety, 42, 1993, pp.327–367.

[4]   McKay, M., *Variance-Based Methods for Assessing Uncertainty Importance in NUREG-1150 Analyses*, LA-UR-96-2695, Los Alamos National Laboratory, 1996, pp.7–11.

[5]   McKay, M., *Evaluating Uncertainty in Stochastic Simulation Models*, Proceedings of SAMO '98, Venice, April 19-22, 1998, pp.171–175.

[6]   Granger Morgan, M. and Henrion, M., *Uncertainty*, Cambridge University Press, 1995, Chapter 8, pp.183–185.

[7]   Sobol, I. M., *Sensitivity Analysis for Nonlinear Mathematical Models*, Mathematical Modeling & computational Experiment, 1, 1993, pp.407–414.

[8]   Saltelli, A., *Are Variance-Based Sensitivity Analysis Methods Healthy For Your Models?*, Proceedings of SAMO '98, Venice, April 19-22, 1998, pp.259–263.

[9]   Saltelli, A., Tarantola, S. and Chan, K.P.-S., *Sensitivity Analysis of Model Output: An Improvement of the FAST Method*, EUR 17338 EN, JRC Ispra, 1997, pp.3–7.

[10] Saltelli, A., Andres, T. and Homma, T., *Sensitivity Analysis of Model Output: An Investigation of New Techniques*, Comput. Statistics and Data analysis, 15, 1993, pp.211–238.

[11] Ang, A. H-S. and Tang, W. H., *Probability Concepts in Engineering Planning and Design*, John Wiley & Sons, 1975, Vol.1, Chapter 4, pp.170–218.

[12] Kaplan, S., *On the Method of Discrete Probability Distributions in Risk and Reliability Calculations – Application to Seismic Risk Assessment*, Risk Analysis, Vol. 1, No. 3, 1981, pp.189–196.

[13] Kaplan, S. and Lin, J.C., *An Improved Condensation Procedure in Discrete Probability Distribution Calculations*, Risk Analysis, Vol. 7, No. 1, 1987, pp.15–19.

[14] Parzen, E., *Stochastic Processes*, Holden-Day Inc., 1962, pp.51–55.

[15] Marseguerra, M., Masini, R., Zio, E., Cojazzi, G., *Variance decomposition-based sensitivity analysis via neural networks*, Reliability Engineering and System Safety,79 , 2003, pp 229-238

[16] Rummelart, D. E., McClelland, J. L., *Parallel distributed processing*, vol. 1, MIT Press, Cambridge, MA, 1986.

[17] NUREG/CR-4769, *Risk Evaluations of Aging Phenomena: The Linear Aging Reliability Model and Its Extension*, US Nuclear Regulatory Commission, April 1987.

[18] Brown, M. and Proschan, F., *Imperfect Repair*, Journal of Applied Probability, 20, (1983), pp. 851-859.

[19] Lim, T.J., *Estimating system reliability with fully masked data under Brown-Proschan imperfect repair model*, Reliab. Engng. and Sys. Safety, 59, (1998), pp. 277-289.

[20] M. Cantoni, M. Marseguerra, E. Zio, *Genetic Algorithms and Monte Carlo Simulation for Optimal Plant Design*, Reliab. Engng. And Sys. Safety, 68, 2000, pp. 29-38.

[21] M. Marseguerra, and E. Zio, *Nonlinear Monte Carlo reliability analysis with biasing towards top event*, Reliab. Engng. and Sys. Safety, 40, (1993), pp. 31-42.

[22] Tarantola, S., *Analyzing the Efficiency of Quantitative Measures of Importance: The Improved FAST*, Proceedings of SAMO '98, Venice, April 19-22, 1998, pp.289–292.

[23] Mosleh, A., Siu, N., Smidts, C. and Lui, C. eds., *Model Uncertainty: Its Characterization and Quantification*, Proceedings of Workshop I on Advanced Topics in Risk and Reliability analysis, Annapolis, Maryland, October 20–22, 1993, NUREG/CP–0138.

[24] Gallegos, D.P. and Bonano, E.J., *Consideration of Uncertainty in the Performance Assessment of Radioactive Waste Disposal from an International Regualtory Perspective, Reliability Engineering & System Safety*, 42, 1993, pp.111–123.

[25] Gallegos, D. P., Pohl, P. I. and Updegraff, C.D., *Preliminary Assessment of the Impact of Conceptual Model Uncertainty on Site Performance, High Level Radioactive Waste Management*,

Proceedings of the Second International Conference, Las Vegas, Nevada, April 29–May 3, 1991.

[26] Evans, J. S. , *Addressing Model Uncertainty in Dose-Response: The Case of Chloroform*, Proceedings of Workshop I on Advanced Topics in Risk and Reliability Analysis, Mosleh, A., Siu, N., Smidts, C. and Lui, C. eds., NUREG/CP-0138, Annapolis, Maryland, October 20–22, 1993.

[27] Eisenberg, N. A., Sagar, B. and Wittmeyer, G. W., *Some Concepts of Model Uncertainty for Performance Assessments of Nuclear Waste Repositories*, Proceedings of Workshop I on Advanced Topics in Risk and Reliability Analysis, A. Mosleh, N. Siu, C.Smidts,C.Lui,eds., NUREG/CP–0138, Annapolis, Maryland, October 20–22, 1993, pp.167–186.

[28] Apostolakis, G., A *Commentary on Model Uncertainty*, Proceedings of Workshop I on Advanced Topics in Risk and Reliability Analysis, Mosleh, A., Siu, N., Smidts, C. and Lui, C. eds., NUREG/CP–0138, Annapolis, Maryland, October 20–22, 1993.

[29] Veneziano, D., *Uncertainty and Expert Opinion in Geologic Hazards*, to be published, 1994.

[30] Zio, E., Apostolakis, G. and Okrent, D., *Towards the Quantitative Treatment of Model Uncertainty in the Performance Assessment of High-Level Radioactive Waste Repositories*, Proceedings of WM'95 Conference, Arizona, USA, February 1995.

[31] Zio, E., Apostolakis, G. and Okrent, D., *On the Use of Expert Opinion in the Assessment of Model Uncertainty in Groundwater Flow*, Proceedings of the Fifth International Conference on Radioactive Waste Manangement and Environmental Remediation, Berlin, Germany, September 1995, pp.739–742

[32] Zio, E. and Apostolakis, G., *A Norm-Based Approach to the Quantification of Model Uncertainty*, Proceedings of the Seventh International High-Level Radioactive Waste Management Conference, Las Vegas, April 1996, pp.252–254.

[33] Zio, E. and Apostolakis, G., *Two Approaches to Model Uncertainty Quantification: A Case Study*, Proceedings of the Third International Conference on Probabilistic Safety Assessment and Management, PSAM III, Crete, Greece, June 1996, pp.631–636.

[34] Zio, E., *Quantitative Analysis of Model Uncertainty: Estimating the Credibility of Alternate Model Structures*, Proceedings of the

International Topical Meeting on Probabilistic Safety Assessment, PSA'96, Park City, Utah, USA, September 1996, pp. 1219–1226.

[35] Zio, E. and Apostolakis, G., *Two Methods for the Structured Assessment of Model Uncertainty by Experts in Performance Assessment of Radioactive Waste Repositories*, Reliability Engineering and System Safety, 54, 1996, pp.225–241.

[36] Zio, E., *Model Uncertainty and the Use of Expert Judgment in the Performance Assessment of a High-Level Radioactive Waste Repository*, Thesis, University of California, Los Angeles, 1995

[37] Siu, N. and Apostolakis, G., *Probabilistic Models of Cable Tray Fires*, Reliability Engineering, 3, 1982, pp.213–227.

[38] Siu, N. and Apostolakis, G., *On the Quantification of Modeling Uncertainties*, Proceedings of the 8th International Conference on Structural Mechanics in Reactor Technology, Brussels, Belgium, Aug. 19-23, 1985.

[39] N. A. Abrahamson, P. G. Somerville, and C. A. Cornell, *Uncertainty in Numerical Strong Motion Predictions*, Proceedings of the Fourth U.S. National Conference on Earthquake Engineering, Palm Springs, California, May 20–24, 1990, Vol. 1, pp.407–416.

[40] Reinert, J.M., Apostolakis, G., *Including Model Uncertainty in Risk-Informed Decision Making*, Annals of Nuclear Energy, 33, 2006, pp. 354-369.

7.11 Appendix

Variance decomposition

Let us denote by capital letters X, Y two random variables and let $f(x,y)$ be their joint pdf. The expected value of the random variable Y can be written as follows:

$$E_Y[Y] = \iint yf(x,y)dxdy = \iint yf(y\mid x)f(x)dxdy = \int \left[\int yf(y\mid x)dy \right] f(x)dx$$
$$= \int E_Y[Y\mid x]f(x)dx = E_X[E_Y(Y\mid x)] \tag{A.1}$$

where:

- $f(y\mid x)dy$ is the probability that Y has assumes a value in dy centred in y when the value of X has been set to x;
- $f(x)dx$ is the marginal probability that X assumes a value in dx centred in x;
- $E_Y[Y\mid x]$ is the expected value of the variable Y conditioned by the value of X being set equal to x; notice that the expected value is computed with respect to Y and as such it remains a function which assumes different values according to the numerical value x of X.
- $E_X[G(x)]$ is the expected value of the variable $G(x)$ function of x.

Similarly, the variance of Y can be written as:

$$Var[Y] = \iint [y - E_Y(Y)]^2 f(x,y)dxdy = \iint [y - E_Y(Y)]^2 f(y\mid x)f(x)dxdy =$$
$$= \int \left[\int [y - E_Y(Y)]^2 f(y\mid x)dy \right] f(x)dx = \int E_Y\{[Y - E_Y(Y)]^2 \mid x\}f(x)dx \tag{A.2}$$
$$= E_X\{E_Y[(Y - E_Y(Y))^2 \mid x]\}$$

Notice that $E_Y(Y)$ is a constant.

Given a random variable Z and a generic constant parameter a one can write:

$$E[(Z-a)^2] = E[(Z-E(Z)+E(Z)-a)^2]$$
$$= E[(Z-E(Z))^2] + E[(E(Z)-a)^2] + 2E[(Z-E(Z))(E(Z)-a)] \qquad \text{(A.3)}$$

Since $E(Z)$ and a are constant, the third term becomes equal to zero and we can write:

$$E[(Z-a)^2] = E[(Z-E(Z))^2] + (E(Z)-a)^2 \qquad \text{(A.4)}$$

In our case, we can write:

$$E_Y[(Y-E_Y(Y))^2 \mid x] = E_Y\{(Y-E_Y(Y\mid x))^2 \mid x\} + [E_Y(Y\mid x) - E_Y(Y)]^2 \qquad \text{(A.5)}$$

Computing the expected value with respect to X and accounting for (A.2), one obtains:

$$E_X\{E_Y[(Y-E_Y(Y))^2 \mid x]\} = Var[Y] = E_X[Var_Y(Y\mid x)] + Var_X[E_Y(Y\mid x)] \qquad \text{(A.6)}$$

This formula is known as *variance decomposition* and expresses the fact that the variance of a random variable Y dependent on another random variable X is given by the sum of the expected value of the conditioned